ROUTLEDGE LIBRARY EDTIONS:
GLOBAL TRANSPORT PLANNING

Volume 14

GERMAN RAILWAYS

GERMAN RAILWAYS
A Study in the Historical Geography of Transport

ROY E. H. MELLOR

Routledge
Taylor & Francis Group
LONDON AND NEW YORK

First published in 1979 by Department of Geography, University of Aberdeen

This edition first published in 2021
by Routledge
4 Park Square, Milton Park, Abingdon, Oxon OX14 4RN

and by Routledge
605 Third Avenue, New York, NY 10017

Routledge is an imprint of the Taylor & Francis Group, an informa business

© 1979 R. E. H. Mellor

All rights reserved. No part of this book may be reprinted or reproduced or utilised in any form or by any electronic, mechanical, or other means, now known or hereafter invented, including photocopying and recording, or in any information storage or retrieval system, without permission in writing from the publishers.

Trademark notice: Product or corporate names may be trademarks or registered trademarks, and are used only for identification and explanation without intent to infringe.

British Library Cataloguing in Publication Data
A catalogue record for this book is available from the British Library

ISBN: 978-0-367-69870-6 (Set)
ISBN: 978-1-00-316032-8 (Set) (ebk)
ISBN: 978-0-367-74442-7 (Volume 14) (hbk)
ISBN: 978-0-367-74447-2 (Volume 14) (pbk)
ISBN: 978-1-00-315789-2 (Volume 14) (ebk)

Publisher's Note
The publisher has gone to great lengths to ensure the quality of this reprint but points out that some imperfections in the original copies may be apparent.

Disclaimer
The publisher has made every effort to trace copyright holders and would welcome correspondence from those they have been unable to trace.

GERMAN RAILWAYS: A STUDY IN THE HISTORICAL GEOGRAPHY OF TRANSPORT

BY

ROY E. H. MELLOR

O'DELL MEMORIAL MONOGRAPH No. 8

(1979)

Department of Geography, University of Aberdeen

ISSN 0141-1454

C O N T E N T S

Page

Preface

Introduction 1

Some Comparisons with Britain 3

The German Railway Network in its Physical Setting 4
    1. *Northern Lowlands* 5
    2. *Central Uplands* 6
    3. *Southwestern Scarps and Vales* 8
    4. *Alpine Foreland and Alps* 9

Historical Growth of the German Railway Network 9

Particularism and the Railways 15

The Aftermath of the Franco-Prussian War 16

Railways and Strategy 18

The Shift from Private to State Railways 19

Particularism and Railway Operations 20

Railways After the 1914-1918 War 21

Creation of the Deutsche Reichsbahn 23

The Years After 1945 24

Aspects of Traffic and Services 31

Freight Traffic 35

****

Tables I-VIII 39

Notes 48

Bibliography 50

LIST OF MAPS

MAP 1 Physical Setting for Railways
2 Major Railway Bridges in Germany
3 Friedrich List's Proposals for an All-German Railway System
4 Growth of Railway Route - I
5 Prussian Railway Connections to the West
6 The Prussian Eastern Railway
7 Growth of Railway Route - II
8 Growth of Railway Route - III
9 The Berlin-Hamburg Railway
10 The Köln-Minden Railway
11 Railways in the Ruhr before State Ownership
12 The Eifelbahn
13 Pre-war Electrification
14 Deutsche Reichsbahn Network in 1937
15 Disruption of Railways by the Post-1945 Inter-German Frontier
16 Berlin Outer Ring Railway
17 Railways of the German Democratic Republic
18 Planned New Routes in West Germany
19 Electrification in West Germany
20 Courier Trains - 1878-1880
21 Fast Passenger Train Services - Summer 1914
22 Express Diesel Railcar Services - Summer 1939
23 Representative Through Coach Services
24 Pre-war and Post-war Railway Passenger Service Densities
25 Inter-City Network in West Germany
26 Germany in the U.I.C. 1973 Plan
27 Passenger Train Services in West Germany
28 Inter-German Railway Services - Summer 1975
29 Inter-war Freight Movements in Germany
30 Container and Piggyback Services in West Germany
31 Freight Services in West Germany
32 Freight Services in the German Democratic Republic

LIST OF TABLES

		Page
TABLE I	The Move to State Railways - The Prussian Example	39
TABLE II	Historical Statistics of German Railways	41
TABLE III	Railway Operations in Principal German States 1885 and 1910	42
TABLE IV	Route Lost after the First World War	43
	Route Lost after the Second World War	43
TABLE V	Railways in the Two German States	44
TABLE VI	Deutsche Bundesbahn	45
TABLE VIIa	Train Services in West Germany	46
b	Express Train Speed in Germany 1914-1938	46
TABLE VIII	The Influence of the First World War on Freight Traffic	47

PREFACE

The O'Dell Monographs were instituted in 1966 in memory of Andrew C. O'Dell, first Professor of Geography in the University of Aberdeen. One of O'Dell's main interests in the broad spectrum of geography he commanded was transport geography. It was, however, the railway that really fascinated him, as reflected in his valuable little book, Railways and Geography (1956), and in his great collection of transport books now housed in Aberdeen University Library.

This short monograph is written in memory of A.C. O'Dell and his interest in the railway. It extends and brings up-to-date his own work on German railways in the Naval Intelligence Handbook, Germany, Vol. IV (1945), for long the best and only review in the English language.

The monograph is by no means exhaustive of the fascinating story of German railways. Though they are perhaps less spectacular than those of Switzerland or less vast than those of North America, they nevertheless provide an interesting study of a railway system and its adjustment to changing political-geographical conditions, let alone changes in economic and social geography. For the transport historical geographer, the comparisons and contrasts with the story in Britain are themselves rewarding.

INTRODUCTION

The growth of the German railway system provides a useful study of the influence of political-geographical factors upon the development of transport systems. Each change in the territorial extent or in the internal territorial-administrative organisation of Germany has had its repercussions upon the spatial pattern of the country's economy and consequently upon the demand for transport. Furthermore, the central position of Germany within the continent has given an added importance to the role of its railways in the overall pattern of the European railway system.

Though the political map of Germany emerged much simplified from the Napoleonic wars, there nevertheless remained some 39 independent territories, many small and poor. Their conflicting interests and jealousies threatened to retard economic and political progress, but the resurgence of commercial life in Europe forced them towards economic unity. This was achieved in the Zollverein of 1834, and thereafter the coming of the railways was to keep the momentum towards political unity going through the widening horizons made possible by the new mobility[1]. Even so, the railways were not to escape from the jealousies and conflicts of the German states, since the different states' postures towards them varied considerably.

The first public railway opened in 1835 and in the following five years several other similar short and local projects were begun. By the mid-1840's, however, this piecemeal construction was beginning to coalesce into longer through routes. Much early impetus came from major commercial towns, like Köln and Leipzig, which quickly realised the implications of the new mobility brought by the railways. The governments of the states were at first suspicious of the railway, but nevertheless several decided at an early date that state participation was preferable to leaving things in private hands, whereas others felt the initiative could remain in private hands so long as they retained some legislative control over railway building and operating. In the particularist atmosphere of the period, the states did not miss the political, economic and strategic significance of the railways and acted accordingly, so commonly influencing the choice of routes and the methods of operating.

Extensive building through the 1850's gradually extended the network and by the mid-1860's the basic system of mainline railways had been largely completed, though important links were still to be added. Having played a key role in the Prussian victory in the war of 1866, the railway was now in a strong position, with the accelerating tempo of industrialisation encouraging further development: a vital instrument in expanding the economic and political horizons that were a major factor in the weakening of parochial, particularist attitudes that culminated in the declaration of the Second Reich under Prussian leadership in 1871. The large indemnity extracted from the French after their defeat by Prussia at this time gave renewed vigour to economic growth, making possible a further stage in the growth of the network, the building of secondary and tertiary routes. The 1870's as a commercially stormy time revealed a growing discontent with the still extensive private railways, notably in the Rhinelands and in Saxony, strengthened by official concern over shortcomings in their strategic performance in the 1870 war with France. The subsequent decade consequently marked the extension and completion of state ownership. Throughout this period, however, the extension of the network continued with little slackening in the pace, though interest began to shift from extending the route to increasing the capacity of its mainlines and junctions (Graph 1).

Everywhere in Europe after the First World War, railways entered a new phase in their existence. In Germany, the years between the two world wars were a time of great economic and political upheaval, arising in no little measure from adjustment to the considerable territorial loss demanded by the peace treaties. On the railways, with the loss of part of the system in the surrendered lands, interest was now on organisational and technological change as their near monopolistic position was increasingly challenged by the rapid growth of road transport. With a radical change in the internal political geography of the Reich, the states' railways (Länderbahnen) were brought under a unified management (Deutsche Reichsbahn) in 1920, in the hope of eradicating soaring deficits by rationalisation and of mounting an effective policy to challenge competing media.

The Second World War was even more destructive than the 1914-1918 conflict: not only were there further extensive territorial losses (largely in the east) and heavy reparations payments, but there was also widespread serious damage to the railways. Most disruptive for the railways has been the emergence of two sovereign states within the remaining territory of Germany, for the boundary between them cuts across one of the most densely developed parts of the pre-1939 railway network, and interchange across it has been at a low level. The integrated system laboriously fostered under the interwar Deutsche Reichsbahn has had to be reorganised into two separate operating systems in a completely new set of spatial patterns of economic and political geography, a change that has had to be undertaken without being able to change fundamentally the network geometry of the system developed to serve the now defunct Reich. Railway development in the two states has also diverged because of the markedly different economic and political philosophies which they have chosen to follow, though in both states railways have also changed technologically, with rationalisation and reduction in their route length in the face of competition from other media.

SOME COMPARISONS WITH BRITAIN

In Britain the demand for massive bulk haulage generated by industrialisation was first satisfied by building an extensive system of canals. By the time the railway had become technologically feasible and was being widely built in the 1840's and 1850's, the main features of the industrial map - the iron-making districts, the great coalfields and the many textile towns - were well established. Not so in Germany: by the time industrial development was beginning to accelerate in the 1860's, the major features of the railway system were already advanced. Demand for freight movement had been the main generative factor in railways in Britain: in Germany, for a considerable initial period, it was passenger traffic that earned the railways their dividends.

The contrast between Britain and Germany is explained quickly by any economic history of the two countries. In Germany, it was only after 1870 that production of coal and iron came to equal and in part ultimately exceed that in Britain. In the same period, there was also a massive growth of population in Germany, when the main population explosion had already passed in Britain. Most striking at this time was the sudden upsurge in urban population in Germany, though the proportion of town dwellers in the total population remained lower than in Britain. Thus it was the years between 1870 and 1914 that made demand for railway transport in Germany comparable to that in Britain. But there was also a spatial difference in Germany - the heavy industrial districts were fewer and more widely scattered on the peripheries than the comparable situation in Britain. No coalfield in Britain, however, held such a dominant position as the Ruhr in Germany, which has consistently contributed at least three-quarters of total coal production and a not dissimilar proportion of iron and steel output. There is also the fact that even though an all-German customs union was achieved in the <u>Zollverein</u>, a strong parochialism and particularism typified the states' railways, whose networks usually showed a focus upon the capital and were in any case designed primarily to serve their own state. Though Berlin became the imperial capital and a powerful force in national political and economic life, it nevertheless did not come to have the same pre-eminence of either London or Paris in their own national situations.

Different commercial philosophies also distinguished Britain from Germany. The British government did little more than regulate railways for safety and reasonably fair trading - private companies were left to build and operate lines, and in this, the government openly encouraged 'competition'. When British railway companies late last century began to make agreements on traffic sharing and other operations, besides suggesting amalgamations, there was no little hostility from government circles, with the vetoing of several proposals. Nationalisation was rejected even in the desperate conditions after the First World War, with four large commercial companies formed in 1923, ostensibly to preserve the element of competition. German public sentiment on the other hand was far less concerned with establishing free commercial operations. Powerful state intervention came early, so that state-owned companies or ones with hefty state participation appeared quickly alongside the private companies. In South Germany, state railways existed from the start, whereas private companies were mostly in the Rhinelands, Saxony and Silesia. In Prussia, both state and private railways existed, but private railways found themselves in financial trouble or, where regarded as strategically important, were readily put under state participation or ownership. No particular effort was made anywhere in Germany to encourage competition between the railways; in fact, it seems more to have been the reverse situation. Where competition appeared to exist, it seems usually more the result of early inter-state rivalry and particularism rather than truly commercial intent. Far less duplication of routes, stations and other installations occurred in Germany than under the inter-company rivalry in Britain.

There were, however, particular German circumstances: for the first 25 years of railway building and at several later stages, there was difficulty in recruiting capital, so that state aid was usually willingly accepted. There was a tendency of private companies only to promote routes believed to be of great financial potential, leaving the states to finance the building of essential but commercially unpromising lines. In a continental situation, railways had, however, a much greater strategic significance than in Britain: after the so-called Railway War of 1866 with Austria and its allies, the Prussian General Staff became intensely interested in railway matters, with generals even suggesting that railway building was more valuable than erecting fortresses. It was a railway war in the sense that much use was made of the new mobility imparted by the railway in strategic and tactical manoeuvre, of which the Germans made better use than the Austrians, so enhancing their superiority in armaments and morale. After the war with France in 1870, Bismarck claimed that the strategic role of railways for the new Reich warranted the incorporation of all state and private systems into one vast Imperial Railway Administration (Reichseisenbahnen), though his dream remained unfulfilled. Nevertheless, when the moment for such a move came in 1920, the new Deutsche Reichsbahn was accepted with less public concern than the railway grouping of 1923 in Britain. In West Germany today there is no feeling in any way equivalent to the views in Britain that British Rail should be split again into constituent companies or at least extensively decentralised.

THE GERMAN RAILWAY NETWORK IN ITS PHYSICAL SETTING

While any railway network serves the spatial pattern of economic geography and population distribution of the region in which it is set, it cannot escape from the influence of the region's physiography: every railway line is influenced by the relief and

drainage of the country it must cross. Each railway network
develops in a particular matrix of economic geography, responding
in various degrees to change in economic and population distribution.
Linkages in the network after a time tend to introduce some inertia
to economic change by holding the most desirable locations along
them. Even so, changing economic conditions may, however, induce
the addition of new linkages or perhaps the discarding of ones no
longer effectively serving needs. Physical conditions nevertheless
influence which are added or discarded: new lines will be built
along the most advantageous trajectory in relation to gradients and
unavoidable civil engineering works, while discarded lines, where a
choice exists, will be the most difficult to maintain and operate.
If a new transport medium emerges, as for example road transport,
the railway network may be ill-placed to serve the changing needs in
the face of a more flexible competitor. In this instance, the
response of the railway is often a re-ordering of existing linkages
into a new hierarchy, where again physiographical conditions along
the routes may be significant in the selection.

Unlike Britain with its poor peripheral uplands in the north
and west, there is a nice distributional balance in Germany between
the richer and more fertile areas supporting industry and flourish-
ing farming and the poorer terrain of both the uplands and lowlands.
Nor is there terrain in Germany as difficult of access to railways
as Upland Wales or Highland Scotland, though considerable physical
obstacles do occur in the uplands of Central Germany. In Germany,
the distribution of the poorer and more sparsely settled country
between richer lands able to generate generous traffics resulted in
mainline railways being built across the poorer terrain in order to
link together such valuable traffic generating districts: the two
comparable British situations are the Scottish Southern Uplands and
the Central and Southern Pennines. This mosaic of richer and
poorer areas, besides the central continental situation, perhaps
also accounts for the absence in Germany of the long, lightly used
deadend branchlines typical of the remoter coastal peripheries and
uplands in Britain.

German territory may be divided into four major physiographic
provinces, in each of which the railway has faced different physical
challenges (Map 1). The later start of railway building in Germany,
soon after which the true adhesive capabilities of the smooth iron
wheel on the smooth iron rail were properly understood, and the
early problems of raising capital compared to Britain, gave less
incentive to construct the beautifully evenly graded trajectories
typical of the early English mainlines. Consequently imposing civil
engineering works are less in evidence though far from absent, while
track is generally laid closely to the configuration of the terrain.

1. Northern Lowlands

The northern lowlands, gently undulating to flat topography,
are characterised by long sections of straight track with few
gradient problems. Construction was therefore relatively inexpen-

sive, an important factor in bringing the railway to the modestly settled, largely agricultural countryside. Where considerable diversions from the obvious direct route occur, they are usually the product of territorial jealousies in the early period when particularist attitudes predominated, though certainly some are to reach significant commercial towns. The broad shallow valleys, including the ancient meltwater channels, provide apparent routeways, but their wet floors that flood easily have usually resulted in railways being built along the drier valley flanks, while such wet conditions pose problems in crossing valleys and long bridges are needed over the broad if shallow rivers. In the northwestern interfluves, railways also faced the difficulties of extensive wet moorland. The rougher young moraine landscapes in the north, with many wet hollows, lakes and hummocky terrain, are reflected in the curving trajectories of most railways, with directness sacrificed for cheapness as most are secondary and tertiary routes built in the latter phases of network development. The many estuaries and lagoons of the Baltic and the major estuaries of the North Sea make railway construction along the coasts difficult, though several important offshore islands have been joined to the railway system by bridges or ferries.

The southern part of the North German Plain, projecting in broad embayments into the flanks of the central uplands, is covered by rich dark soils and often underlain by coal, lignite and even iron ore. This economic richness is marked by the notably denser railway network than general in the northern lowlands, though construction problems were similar to further north. In places, some railways were laid on rather circuitous routes to avoid crossing outliers of the central uplands, as well seen in southern Lower Saxony.

2. Central Uplands

The roughly dissected upland country in Central Germany forms the second major physiographic unit. Railway builders were presented not only with considerable relief obstacles but were also burdened by the jealousies of the many small states in this diverse and fragmented landscape. Fortunately, numerous well-defined routeways along river valleys between the blocklike uplands exist, though easy east-west routes are less common than those running north-south. The principal constructional problems arose generally where routes were forced to cross from one valley system to another, though expectedly much use of saddles and passes has been made. It is perhaps not surprising that the majority of German tunnels occurs in these uplands and includes some drilled early this century when shortcuts to otherwise circuitous routes were being made: similarly, such terrain has necessitated innumerable bridges and viaducts.

The most impressive feature of these central uplands is the line of blocks and ridges running from the Teutoburger Wald in the northwest to the Thüringer Wald and Czech-German border country in the southeast, a substantial barrier to easy railway links from

North Germany to the south and southwest. The northwestern ridge-like scarps of the Teutoburger Wald and Wiehengebirge are reflected by the sinuous and roughly graded section of the Bremen-Münster line between Bohmte and Lengerich; the Hannover-Hamm line uses the Weser Gap (Porta Westfalica) at Minden and a low saddle at Bielefeld to cross the eastern end of this section. Further southeastwards, the Egge presents a sharp ridge, crossed by use of the main saddle, the Neuenheerse, while a tunnel under the Rehberg brought the growth of Altenbeken as a railway town at the expense of Paderborn. The confused pattern of vales between ridges of various shapes in the Weser Bergland is reflected in the equally confusing pattern of railways seeking routes through it. The Weser Valley, meandering and narrow, is not used as a continuous railway route, since the broader and less meandering Leine Valley has been preferred as the more direct north-south link. On the south, the Fulda Gap at Kassel and the Eichenberg Pass on the Werra provide gateways to the Hessian Corridor through the uplands to the Rhine-Main confluence. A formidable barrier is the massive, dissected Thüringer Wald and its southern continuation in the Frankenwald and Fichtelgebirge. Here deeply incised saddles across the uplands are absent and even valley approaches are poor, so railways must climb almost to the summit level, avoided, however, by the 3 km. long Brandleite Tunnel on the Erfurt-Suhl line. A major railway follows the upper Saale and the Loquitz through the Lauenstein Gap to the upper Hasslach that leads to the Main (Halle/Saale-Bamberg-Nürnberg), while further east, the Leipzig-Gutenfürst-Nürnberg line crosses the roughly dissected but lower Vogtland and picks its way through the Fichtelgebirge, a route marked by numerous imposing civil engineering structures. The formidable upland along the Czech-German border has only three railway crossings, of which the easy route through the Egerland Gate (Schirnding-Cheb) is the principal one.

Although the Hessian Corridor, with its low dissected plateaus and basins, provides an easy railway route from the upper Weser to the Rhine-Main confluence, the southern end is straddled by the massive low basaltic dome of the Vogelsberg. On its eastern flank the railway from the north bifurcates to Frankfurt and to Würzburg and the original junction necessitated trains between Frankfurt and the north reversing in the station at Elm, until eliminated by a tunnel in 1916.

The northern German-Czech borderland, including former German Silesia, also shows a pattern of railways following convenient valleys and troughs, seeking low passes through formidable uplands. Railways across the Erzgebirge are all of secondary importance: several follow the gentle gradient of the Saxon face, but few climb the steep scarp on the Czech side from the Ohře (Eger) Valley. The main routes from Germany into Northern Bohemia are, on the west, through the Elstergebirge and, on the east, through the Elbe gorge between Bad Schandau and Děčín. East of the Elbe, the upland blocks separated by broad depressions and valleys were marked by the growth of a considerable railway network before 1914, even though the Riesengebirge and Altvatergebirge were left virtually without railways. Steep gradients in the Silesian lands led, however, to early electrification of several railways around Görlitz and Hirschberg (Jelenia Góra).

The influence of valley routes upon railways is well illustrated in the Rhine basin. The Rhine itself is a serious barrier: in the 694 km. of the German reaches there are only 17 railway bridges (additionally five not replaced after damage in the 1939-1945 war) (Map 2). Railways extend, however, along the entire length of both banks, particularly impressive in the narrow gorge sections from Bonn to Andernach and from Koblenz to Bingen. Interestingly, the Rhine-Saar link follows not the narrow, meandering Mosel but the straighter trough provided by the old pre-Tertiary valley (Moseltrog). In the Sauerland, the industrial Siegerland is reached from the north by a railway, with many substantial civil engineering structures, along the narrow Lenne Valley, while the constricted Sieg Valley is used for railway access from the Rhine. In the northern part of the Sauerland, the valleys of the Ruhr and Diemel provide a route eastwards to Kassel. The Lahn Valley, nowadays a secondary route, gives access to Wetzlar. West of the Rhine, the Erft and Kyll Valleys are used for the steeply graded route across the western Eifel from Köln to Trier, with a bleak and exposed summit section, built to strengthen strategic ties with the Saar and Lorraine after 1871. Some secondary lines cross the rough southern scarp of the Rhenish Uplands in the Taunus; but the higher ridgelike Hünsruck on the west has only one longitudinal railway of minor importance.

3. Southwestern Scarps and Vales

The third physiographic region comprises the scarp and vale landscapes of Southwest Germany. It is fortunate for railways that the scarps have all been deeply scarred by streams that provide easy access to their summits. The major river valleys draining this region have also been used for railways, though seldom throughout their whole length, but the relationship between the railway network and the drainage pattern is nevertheless close. The imposing scarp of the Suabian-Franconian Jura (or Alb) has formed the principal obstacle to railway access out of the region on the south and east, though a complex interplay of economic and political factors resulted in some of the most obvious routes not being used or developing only a secondary role. The major route across the Alb is from Stuttgart to Ulm (-München), which after much deliberation, was built along the Fils Valley and over the 1:40 ramp of the Geislinger Steige, the most direct but not the easiest route. The most obvious route further east across the Ries basin was not chosen for the mainline railways, which took a more circuitous trajectory via Treuchtlingen, with one through the Altmühl gorge and another across the open dip slope.

A notable obstacle between the broad flat floor of the Rhine rift valley and the interior scarp and vale landscapes is the Schwarzwald, but as this was also the boundary between Baden and Württemberg, no great incentive existed originally to build railways across it. Ultimately, the important Offenburg-Donaueschingen rail-

way from the Rhine to the Bodensee followed for much of its length
the easy valley of the Kinzig, crossing the summit to the upper
tributaries of the Danube by means of reverse loops to ease the
descent, while the Höllentalbahn, from Freiburg im Breisgau to the
upper Danube, demonstrates a high level of engineering skill, with
clever use of viaducts and positioning of the line to minimise the
likely closure by snow.

4. Alpine Foreland and Alps

The fourth physiographic province comprises the Alpine Foreland
and the narrow belt of the German Alps. The broad undulating Foreland, broken in a few places by rolling hill country, presents no
serious obstacles to railway building, with the main difficulty in
crossing the wide, braided streams draining north to the Danube,
and some peaty fens along the latter's course. Along the mountain
foot, however, young moraines and rocky ridges influence railway
trajectories (e.g. as at Murnau). The railways tend to follow the
main rivers that provide access into the mountains, like the Inn
Valley above Rosenheim, though the direct route from München to
Innsbruck, across rougher terrain, uses the Würm, Ammer and Loisach
Valleys, the narrow Scharnitzer Klause (Porta Claudia) and the difficult Seefelder Sattel.

HISTORICAL GROWTH OF THE GERMAN RAILWAY NETWORK

German miners used simple trolleys on wooden rails in the
sixteenth century, while a cast iron plateway was laid at a mine in
Clausthal (Harz) about 1775. In 1815, a locomotive was built in
Berlin for coal haulage at Königshütte in Upper Silesia, but it was
the wrong gauge and its ultimate fate remains a mystery. In 1827,
the far-sighted industrialist Friedrich Harkort built a railway at
his Muttental works near Witten on the Ruhr, where several plate-
ways already existed, serving local coalmines. It was, however, the
success of the Stockton and Darlington (1825) and the Liverpool and
Manchester (1830) railways in England that really stimulated German
interest, although as early as 1807, a Bavarian minister, having
seen mine tramways in England, envisaged a "road with ironways"
between the Rhine and the Danube[2]. In 1828, the Prussian Minister
Motz suggested a railway between the Lippe (Lippstadt) and the Weser
(Minden), a demand also made in 1831 by the Westphalian Landtag[3].
The latter, supported by Harkort, in 1833, proposed a railway from
Köln to Minden, while in Köln businessmen began to press for a railway to Antwerp to avoid paying the high tolls levied by the Dutch on
Rhine traffic passing out to sea. All these early proposals were
in a sense 'portage' lines between main river basins and aimed to
give early access to ports. Fired by experience in America, the
most ambitious project was put forward in 1833 by the economist

Friedrich List[4]. This was to be an All-German railway system (Map 3) centred on Saxon Leipzig and radiating to all the major German towns. For such vision, public opinion was not ripe and the project was not instantly taken up, though ultimately nearly all the suggested links were built, even if in a less coherent way. It is interesting that France, a more centrally managed state than the congery of German princedoms, quickly accepted the idea of a national railway system, for which Stephenson in England had also made a case, although unsuccessfully.

The first true railway was not one of these more ambitious schemes but a modest five-kilometre line between the adjacent towns of Nürnberg and Fürth. In 1833, a forty-day traffic survey had shown an average daily movement of 1,184 people on foot, 494 people in 185 carriages, and 108 wagons drawn by 236 horses between the two towns. Seeking advice from 'English specialists', the promoters had the Bavarian Land Survey draw up a level trajectory between the towns. An English locomotive (a Stephenson 2-2-2) was bought, but the rails and chairs came from the Rasselstein ironworks at Neuwied since the Bavarian government refused to allow duty free import from England. As an aside on contemporary transport, the locomotive (packed in 19 crates weighing 170 cwt.) left Newcastle on 27 August 1835 and docked at Rotterdam on 17 September. On 7 October, it arrived by river boat at Köln and was loaded on to eight horse wagons, to reach Offenbach by 15 October. Here, the drivers refused to go further because of the state of their wagons' wheels and axles, so it was reloaded on to other wagons, ultimately to arrive in Nürnberg on 28 October. An English driver and mechanic erected it and then trained local men: the first steam-hauled train ran on 16 November and on the opening day (7 December 1835) the locomotive drew a train of 200 passengers along the line in nine minutes.

Preparations by this time were already well advanced for a more ambitious project to join Dresden, the Saxon capital, to the great commercial town of Leipzig. Without a good river route, the latter town was gravely concerned about its future in the upsurge of economic life in Europe and the railway was seen as a way out of its problem. An English consultant advised against List's proposed route, because of the cost and difficulty of passing through the well-populated Mulde Valley and the rougher country towards Meissen, so instead the line was built through open country via Riesa. A constraint on the route chosen was that a gradient greater than 1:300 was then considered beyond the adhesive capabilities of locomotives. The first four miles, from Leipzig to Althen, opened in 1837, included the first German tunnel, 512 m. long and dug by 380 Freiberg miners. The line also demanded substantial bridges at Riesa and Röderau and a deep cutting at Machern, but it was an immediate success and carried over 400,000 passengers in its first year.

By the time the whole 115 km. were open between Leipzig and Dresden in 1839, railways also joined Braunschweig to Wolfenbüttel (the first state railway), Berlin to Potsdam (the first Prussian railway), while the Düsseldorf-Erkrath line, including a rope hauled incline[5], was part of the railway to Elberfeld to tap the prosperous textile district. In 1840, the Leipzig-Halle-Magdeburg railway opened. In the south, Baden was building its mainline from Mannheim to Basel, opening the first part from Mannheim to Heidelberg in 1840. Unlike the other railways on the English 4' 8½" gauge (1,435 mm.), it was being built at the 5' 3" gauge (1,600 mm.) and in all 485 km. of route was laid before conversion to standard gauge in 1854-1855, after Baden failed to interest its neighbours in broad gauge railways. By 1841, the first part of the Köln to Antwerp railway opened as far as Aachen, the Schelde port being reached in 1843 with Belgian financial assistance.

The contemporary mind in Germany had boggled at the implications of large scale railway planning in organisational, technical and financial terms, but several states quickly appreciated the need for broad legal regulation of railways, especially as piecemeal construction of individual railways began to fuse into longer routes and, even more significantly, began to cross state boundaries. One of the most far-sighted was Prussia, whose Railway Law of 1838 laid a basis for much later legislation. Saxony, whose nodal position was to benefit it in the railway age, went for state railways and a progressive policy of building railways to encourage transit traffic across its borders. Hannover, without large industrial capital resources, also built state railways, as did Braunschweig initially. Baden was early committed to state railways, like Bavaria, though the wealthier Bavarian Rhenish Palatinate had railways developed by private capital. Even though states agreed on operating regulations for railways, there was often long haggling before lines between them were built. The Free City of Frankfurt sought to recapture some of its lost importance by promoting railways to its main trading partners, but found that the confusing territorial pattern of the Hessen states around it resulted in drawn out negotiations, lasting from three years over the Frankfurt-Wiesbaden line to fourteen years for the Frankfurt-Homburg railway. In other cases it was personalities that caused procrastination: Ludwig of Bavaria preferred the canal faction to the railway supporters, whereas the ruler of Hessen-Kassel by his indolence marred the grandiose plans of Kassel to become a major railway junction.

Within five years of the opening of the first railway, some 500 km. of route in ten sections had been opened and many further sections were being built, so that by 1845 the route length had risen to over 2,000 km. (Map 4), three-quarters run by private companies and the rest by the rapidly growing state railways. As speculation in railway building grew and something of the railway mania seen in Britain threatened, the states took action to control and outlaw speculative building, so commonly increasing their own involvement. Prussia in particular, with a sounder financial position every year, began to indulge in building or in financing and guaranteeing construction. It was encouraged in this by Nagler, the chief postmaster, who changed from the railways' bitterest enemy to their most enthusiastic supporter, proposing several routes on his own initiative. Map 4 shows that by 1845 a journey from Stettin via Berlin was possible to Hannover or (via Leipzig) to Zwickau or Dresden.

Two years later, from Dresden one could continue to Breslau and to
the eastern border at Myslowitz. The opening of the Köln-Minden
railway in 1847 made a railway journey from Berlin via Minden and
Aachen to Antwerp possible, and railway trains already (1846) joined
Berlin to Hamburg, while the same year Bremen was linked to the
capital. In the south, much of the Baden mainline was complete,
with trains running from Mannheim to Freiburg, and in Bavaria, the
railway from Plauen and Bamberg via Nürnberg was joined to München
by 1849.

As the various companies' tracks extended and began to join up,
the network of through routes spread, so that by 1850 the railway
map was already beginning to show the main characteristics of the
modern system. Five lines already radiated from Berlin; the
Saxon lands were served by an appreciable basic network; while the
Rhenish-Westphalian and Rhine-Main areas were emerging as focal
points of local systems. Within the next five years, almost all
the principal provincial towns were to be drawn into the railway
network as additional through routes were completed, but there was
commonly much debate as to how far routes should deviate from the
most direct connection in order to serve significant communities
along them: for this there appears to have been no clearly expressed
procedure. For example, the completion of the line from Frankfurt
via Giessen and Marburg to Kassel made the through journey from
Berlin to Freiburg im Breisgau possible in 1852, while a year
earlier the building of the Reichenbach-Plauen line completed the
Berlin-Leipzig-Nürnberg-München route. This last section included
the massive Göltzschtal viaduct, still the largest of its type in
the world. It was indeed sometimes quite short lines that made
significant contributions to the overall system; for instance, the
little Prussian-owned Saarbrücken Railway in 1852 joined the railways
of the Bavarian Palatinate at Bexbach to those of France at Forbach,
a distance of about 30 km., so creating an international route of
significance.

Railway building was early influenced by Prussian attempts to
join firmly together its eastern and western provinces. Private
railways in Prussia and the state railways of Hannover and Braunschweig
first forged this link in 1847. Deteriorating relations with Hannover
and a proposal by Nagler for a line through states friendly to Prussia
were followed by a second route in 1853 through Erfurt to Kassel and
then through Warburg to Hamm. In 1865, just before the Prussian
victory over Hannover, a third route, even more extensively across
Prussian territory was opened. This ran from Oschersleben via the
Braunschweig Railway to Kreiensen and then on to Höxter, Altenbeken
and Paderborn to Hamm (Map 5). Prussian desire to have easier
access to its Silesian arsenal also underlay the completion of a
railway from Berlin via Frankfurt an der Oder to Breslau in 1846,
which in 1848 was extended via Ratibor and Oderberg to form a
through route to Vienna. The importance of this latter connection
was eclipsed, however, in 1851 by completion of a more direct Berlin-
Vienna route via Dresden, Prague and Brno. 1851 was also marked as
the year in which all the Berlin terminal stations were first joined
together. A strategically important railway opened in 1853 with

completion of the Berlin-Königsberg line, except for the bridge over the Vistula at Dirschau (Tczew), first completed in 1857. The original roundabout route was shortened in 1857 to run via Kreuz and Küstrin instead of Stettin, while in 1867 the Berlin-Küstrin cut-off was built and in 1873 a direct route from Schneidemühl via Konitz to Dirschau was opened (Map 6).

The rougher terrain of South Germany raised constructional costs in an area where capital was not easy to raise; but this was offset by the positive attitude towards railways of the larger South German states, prepared to finance construction rather than wait for private enterprise. By the early 1840's Württemberg had decided to construct state railways and the first section opened in 1845, part of the main line from north to south, completed in 1853. The important but steeply graded route (1:40) across the Geislinger Steige to Ulm was opened by 1850, after several alternative routes across the Suabian Alb had been proposed. The route finally chosen was a German proposal, though the British engineer Vignoles had first been consulted. Württemberg railway practice also came to be strongly influenced by American ideas, whereas elsewhere in Germany English influence prevailed. Reference has already been made to Baden's broad gauge policy that delayed its links to adjacent states, and it was only in 1855 that Basel had been reached across the difficult section over a spur of the Black Forest at Istein. Few railways came to be built across the Black Forest, and then mostly in later times after the early particularist feelings between Baden and Württemberg had weakened. These feelings were also responsible for the failure to create an effective mainline link between Stuttgart in Württemberg and either Mannheim or Karlsruhe in Baden, a weakness that remains in the Southwest German network to this day. Württemberg was equally unenthusiastic in providing a link to the Bavarian railways at Ulm, so that not until 1854 had the Ulm-Neu Ulm line completed the Stuttgart-Augsburg-München route, and only in 1858 was access to the Main Valley from the Rhine made by the opening of the Darmstadt-Aschaffenburg railway.

Completion of the railway along the west (left) bank of the Rhine was an important event of the 1850's. Though the Köln-Bonn railway opened in 1844, it was not until 1859 that the remainder of the line to Mainz was finished, giving access to the railways of the Palatinate and of the Saar. The right bank line, in contrast, was not completed until 1871. There were, nevertheless, few crossings of the river - the first was the train ferry at Duisburg from 1848 until replaced by a bridge in 1866. The most used crossing was to be, however, the Dombrücke in Köln, opened in 1859 at the same time as the 'Centralbahnhof'. Thereafter bridges opened at Mainz (1862), Koblenz (1864) and Mannheim (1867), but ferries were also run (Griethausen by Elten, 1865, and Bonn, 1870). The bridge of boats built at Maxau in 1865 was not replaced until 1938.

The 1850's also saw some useful railways completed in the Prussian eastern provinces, like the Posen-Breslau line (1856) that gave direct access from Silesia to the Baltic port of Stettin. By the end of the decade, however, the pace of railway building in the eastern provinces was losing momentum, despite the powerful Junker lobby and even strategic considerations, to that in the central and western parts where industrialisation was accelerating and increasing the demand for railway transport, besides generating capital for construction. Two typical completions of this period were the opening of a direct railway from Berlin to Leipzig via Bitterfeld in Central Germany (1859) and opening of railways from Münster and Osnabrück to Emden via Rheine (1854-56), giving Westphalia better access to the sea.

The railway map of 1855 (Map 7) with 8,500 km. of route, had already shown a journey across Germany from Aachen on the west to the Russian frontier in Upper Silesia on the east was possible, though the Rhine crossing had to be made by boat until the opening of a railway bridge at Köln in 1859. From Königsberg, a railway journey could now be made via Berlin to Lindau or Friedrichshafen on the Bodensee, while from Flensburg, allowing for a walk between Altona and Hamburg, the traveller could go by rail from the far north, albeit rather circuitously, to Basel. The following year, a more direct route opened from Harburg (opposite Hamburg on the south bank of the Elbe) to Basel or München via the Göttingen-Kassel railway. Though the main railway focus in Northwest Germany, the several railways serving Hamburg were not joined together - from Altona, trains left for Kiel and Flensburg (both like Altona, then in Danish hands); from Hamburg, Berlin was the main destination, while the Harburg trains ran to Hannover. Hamburg and Altona were only joined in 1865 (the line running along streets between the two towns) and not until 1872 was a bridge across the Elbe to Harburg opened.

The 1860's opened with the junction of German and Russian railways at Stallupönen in East Prussia, while in the far south, the final link in the München-Salzburg-Vienna route was made. This augmented the München-Kufstein-Innsbruck route to Austria of 1858 and was quickly followed by completion of the Nürnberg-Passau-Vienna route. The decade was one of feverish growth of heavy industry in the Rhenish-Westphalian area, with consequent growth of the local railway system, of which the major railway completed was the east-west mainline from Dortmund-Bochum-Essen-Mülheim to Duisburg and Oberhausen (Berg and Mark Railway) in the years 1860-1862. Elsewhere around 1863 the strategic Lahn route from Wetzlar to Oberlahnstein opened, and the useful lines from Bremen to Bremerhaven and from Berlin and Stettin via Anklam to Stralsund were finished. In 1867, easier access from Berlin to Silesia came with the opening of the line through Görlitz, while by building short missing sections, the Bebra-Fulda-Hanau line opened in 1868 and the Würzburg-Ingolstadt-München route between 1867-1870. The emphasis had now shifted from major trunk routes to completing other mainlines and to establishing shorter and more direct routes, like the improved Karlsruhe-Stuttgart connection of 1863. By 1870 few mainlines remained to be added (Map 8).

PARTICULARISM AND THE RAILWAYS

The evolution of the network was influenced from early days by the particularist attitudes of the German states. The Berlin-Hamburg Railway set the pace in diplomacy in the mid-1840's (Map 9). Hannover would not let this Prussian line across its territory unless the terminal was in Hannover's Harburg south of the Elbe and not in the Free City of Hamburg, a condition unacceptable to the Prussian and Hamburg sponsors. Even the more northerly route consequently chosen was forced further northwards than necessary to satisfy Mecklenburg's condition that in crossing its territory it must be possible to build a line to the capital, Schwerin. The line thus ran via Hagenow. The line also had to run across the Danish-held Duchy of Lauenburg, allowed only if all traffic between Lauenburg town and the mainline at Büchen was carried on the branchline free of charge, the so-called Lauenburger Privileg that lasted until 1935.

Reference to Prussian attempts to join its eastern and western provinces together by railway has already been made. Prussia had originally sought a northern route across Hannover through Neustadt am Rübenberge in order that a line might conveniently be laid to Nienburg and Bremen to join Prussian Westphalia to a seaport. Hannover had, however, insisted on the longer and less convenient route through Minden to maximise the distance (and consequently the financial benefits) across its own territory, opened in 1847. The longer route via Erfurt and Kassel, thought up by Postmaster Nagler, avoided Hannover completely and opened in 1853. With goodwill from Braunschweig, a shorter route still, via Kreiensen and Holzminden, was finished in 1865, but only at the expense of crossing rough country with formidable gradients. This had purposefully avoided the important commercial town of Einbeck because it was in Hannover. In its turn, Hannover had chosen to build its Southern Railway to Kassel across its own bleak Dransfeld Plateau rather than through the easier route of the Eichenberg Pass in Hessen.

In the north, the Danish king prevented a direct link from Lübeck to Hamburg, because of the latter's poor relations with Denmark. Danish opposition generally to railways between the Elbe and Baltic arose from fear of a decline in shipping through the Danish waters in the Sound, with loss of trade for Copenhagen. The Hamburg-Lübeck railway was quickly completed after Prussia annexed Holstein in 1865.

In South Germany, Bavaria had chosen in 1853 to build a railway to its Bodensee port, Lindau, across its own panhandle of rough and difficult terrain rather than cross into Württemberg, while Württemberg built its line to Friedrichshafen as direct competition. In 1869, Baden built the expensive Schwarzwaldbahn (Offenburg-Donaueschingen) to give a shorter approach from the Rhine to the Bodensee without crossing Swiss territory near Basel and Schaffhausen; but in so doing, at Schramberg near Triberg, it avoided running over a fragment of Württemberg's territory by carrying the railway 100 m.

higher than otherwise necessary. In some instances, it was individual towns that suffered, as in Baden, where the main north-south railway avoided Lahr, because its liberal ideas were despised by the government. The Main-Neckar Railway from Frankfurt went to Friedrichsfeld, a site in open country, rather than directly to either Mannheim or Heidelberg, again because of local disagreement over which town should be the terminus of the railway[6].

THE AFTERMATH OF THE FRANCO-PRUSSIAN WAR

The Franco-Prussian War of 1870 and the creation of the Second Reich were to undermine many particularist stands. This was greatly to strengthen the position of the railways, demonstrating even more clearly than the war of 1866 their strategic importance, while it also generated economic growth that brought a rising demand for transport. For a time, it was the western frontier districts that saw important new railways, like the Köln-Gerolstein-Trier line across difficult country with heavy gradients in the Eifel (1871). The direct Trier-Koblenz line along the Mosel trough, opened in 1879, included Germany's longest tunnel, over 4 km. long, at Cochem. Both these routes drew the Saar and the newly acquired territories in Lorraine more closely into the Reich. Closer ties between the Hansa ports and the Ruhr and Lower Rhine were created by piecemeal completion of the railway to Hamburg - the section Wanne-Haltern-Münster opened in 1870, from Münster to Osnabrück in 1871, Osnabrück-Bremen in 1873 and from Bremen to Hamburg in 1874. The line was built by the Köln-Minden Railway (Map 10), which had taken over the long-standing but abortive concept of the Hamburg-Venlo-Paris Railway. While the Haltern-Wesel-Geldern-Venlo section was also completed, it was never of more than local importance. Another abortive railway was a private Dutch-German consortium (The North German-Brabant Railway), opened in 1878, which brought traffic from Flushing to Boxtel in Holland and then via Goch and Xanten to the Rhine bridge at Wesel. Until the 1914-1918 War it provided the fastest route from England to North Germany and Berlin, but after 1918 it was squeezed out of existence by the Dutch and German state railways. With capital easy to recruit, the 1870's marked the building of several direct routes between major towns across country with little traffic generating potential - examples are the fast strategic route from Berlin to Lehrte near Hannover via Oebisfelde; the Berlin-Dresden Railway (via Elsterwerda) opened throughout its 180 km. in 1875; and the Berlin Northern Railway via Neubrandenburg to Stralsund (223 km., opened 1877-1878).

The economic upsurge after the Franco-Prussian War was notably reflected in the growth of goods traffic, particularly from the rapidly expanding Ruhr coalfield (Map 11). Here the Köln-Minden and Berg and Mark Railways held most of the traffic, but in 1878 the Westphalian Railway tried to corner more for itself by building a primarily freight line from Welver to Dortmund and then along the Emscher Valley through Herne, Horst, Bottrop, to Sterkrade. The line was virtually parallel to an existing route of the Berg and Mark Railway's tracks, whose tactics probably contributed to the fact that traffic proved insufficient, to the disappointment of the Westphalian Railway. Some sections were closed after a short life (Bodelschwingh-Dortmund (1882) and Horst-Sterkrade (1884)), but those on the roadbed of existing tracks were re-opened by the early 1900's (e.g. Osterfeld Süd-Horst-Bismarck). Otherwise in the Ruhr numerous short but well used railways were built to serve new industrial and mining sites. Industrial growth in Saxony also resulted in building of short lines primarily to serve small industrial towns, and thus expectedly most of these new lines were in the triangle Leipzig-Chemnitz-Glauchau and included the Zwickau district. In the south, this was rough country and even short lines could demand expensive engineering works, though the same was true of new lines in the hills south of the Ruhr, where the Müngsten viaduct is the classic structure, spanning the Wupper in a single arch, 107 m. high.

After about 1875 new construction had become essentially completion of secondary routes, some of considerable length like the Dortmund-Gronau-Enschede line (1878) or the line from Oberhausen via Coesfeld and Burgsteinfurt to Rheine (1879). Building of such cross-country routes was typical of the Prussian northern and eastern provinces, with construction executed as cheaply as possible - like the Mlawkaer-Marienburg Railway, 149 km. long, opened in 1876-1884, or the 127 km. long line from Warnemünde to Neustrelitz financed by the Norddeutscher-Lloyd shipping company (1886), which ran passenger vessels from Warnemünde to Denmark. The mid-1870's also saw much building in the northwest, notably in Friesland, but an Anglo-Dutch railway project to tap Baltic traffic by linking Dutch Harlingen via Oldenburg to Lübeck, first mooted in the 1840's, never materialised. In fact, late last century several English financial interests in North Germany were bought out under Prussian pressure.

Between 1865-1875 some 13,200 km. of route were added, including 840 km. in the annexed territories of Alsace and Lorraine. Up to the outbreak of war in 1914, a further 4,000 km. of mainline and 22,000 km. of secondary route were laid. Increasingly the laying of new track was absorbed by doubling or quadrupling of existing lines, though it is difficult to demonstrate this from contemporary statistics. After 1892, the passing of a Light Railway Law in Prussia gave a new incentive to railway building and other states enacted similar laws. In Saxony and the Lower Rhine, such light railways included a considerable length of narrow gauge route, mostly in short local systems. Nevertheless, these laws did not result in the building of extensive 'vicinal' railways as happened in Belgium and France.

RAILWAYS AND STRATEGY

Though the Eisenbahnkrieg of 1866 between Prussia and Austria first alerted attention to the strategic potential of the railways, the Franco-Prussian War revealed some serious shortcomings. It was therefore not surprising that the Imperian General Staff, when formed, began to take a keen interest in railway matters, approving and encouraging the building of such strategic routes as Köln-Gerolstein-Trier and Berlin-Oebisfelde-Lehrte. It also exerted pressure for the development of a strictly military route from Central Germany to the Rhine: this was to be part of the so-called Kanonenstrasse between the eastern and western frontiers of the Reich. East of Berlin the existing routes were considered adequate, but to the west a new route was created by building a few short missing sections, work already begun by Prussia before the 1870 War. From Blankenheim on the Harz flank, the route ran via Leinefelde, Treysa and the Lahn Valley to the Rhine near the great fortress of Ehrenbreitstein at Koblenz. It included a short avoiding line (Wetzlar-Lollar) around Giessen in Hessen, so that the line lay entirely across Prussian territory. This route never appeared as a through route in timetables and much of its length only carried local passenger and freight traffic, though it was built and maintained to high capacity mainline standard, and eventually extended from Koblenz via Trier to Metz. Another strategic line begun but never completed ran along the west of the Rhine from the Krefeld-Neuss area through Weilerswist and Rheinbach to the Ahr Valley near Ahrweiler. It was also intended to provide a direct, fast link from the western Ruhr to the Saar and Lorraine via the Ahr Valley and Gerolstein. Work was finally abandoned only in 1923. Effort was also made to improve the crossing of the Rhine, with the bridge at Remagen in 1913 replacing the Bonn ferry and the Engers bridge opened in 1916, while three strategic bridges were built on the Upper Rhine in 1878 to give easier access to the critical Belfort Gap. The Prussian military also forced the building of an outer ring railway around Berlin to ease troop movement, though the project was never completed in its day. Despite strong military influence on general railway matters, Prussia nevertheless decided that special railway troops must be created to avoid some of the pitfalls of the situation in the war of 1870. Consequently a purely military railway was built from Berlin-Schöneberg via Zossen to Jüterbog, parallel over much of the way to the Berlin-Dresden Railway. It was opened in 1875-1897, often being used for experimental rolling stock.

In South Germany, the Wutachtalbahn from Waldshut via Blumberg to Immendingen was seen as a strategic line avoiding Swiss territory at Schaffhausen, besides augmenting the relatively modest capacity of the Höllentalbahn. It ran across very awkward terrain, but it was built generously, with wide curves and gentle gradients as well as a roadbed for double track. After its opening in 1890, however, it carried only modest local traffic. While such lines are generally recognised as having been primarily strategically motivated, all round the frontiers numerous rural railways of potential military importance were constructed. This was most striking in the Prussian eastern provinces, where the density of the railway system was a contrast to the sparse network in Russian Poland, a juxtaposition still seen in the railway system of modern Poland.

While through the 1870's short sections of line were being built to improve through routes, the outstanding possibilities were becoming more costly, involving more daunting engineering work. Among examples are the direct route across the Thüringer Wald between Plaue and Suhl (some 33 km., completed in 1884 with the opening of the 3 km. long Brandleite Tunnel); the short cut at Dillenburg on the Siegen-Wetzlar line by means of an imposing tunnel under the Kalte Eiche (1913); and the elimination of the awkward station arrangement at Elm on the Frankfurt-Fulda line by building the 3 km. long Distelrasen Tunnel (1916). Of a different type, the improvement in railway links to Scandinavia by provision of costly train ferries and harbours included the Warnemünde-Gedser ferry of 1903 and the Sassnitz-Trelleborg ferry of 1909. This direct link to Sweden included the problem of banking on the steep ramp down to the Sassnitz ferry terminal and the train ferry across the Strelasund - the latter not replaced by a bridge to the Island of Rügen until 1935.

THE SHIFT FROM PRIVATE TO STATE RAILWAYS

In the turbulent commercial times after the Franco-Prussian War, several private railways in financial difficulties had to accept state aid to put them on a sound footing, usually resulting in their takeover by the state railways. Beset by these problems and the official discontent over the railways' performance in the war, Prussia enacted a law in 1876 to take over all private railways before 1887 (the task was largely achieved, however, by 1886 (Table I)). The same year, Saxony bought the important Leipzig-Dresden Railway and completed state ownership by 1882. The process began in Bavaria at the same time, with the purchase by the state of the extensive Bavarian Eastern Railway, though several small railways were nationalised only around 1906-1910 and the extensive Bavarian Palatinate Railway was only taken into state ownership in 1908. A rather exceptional situation had been the sale by the state in Braunschweig of its railways in 1870 to the Darmstädter Bank, which sold out again in 1885 to the Prussian State Railways. In 1896, the amalgamation of the extensive Hessen Railways with those of the Prussian State system came to dominate the German railway scene, with almost two-thirds of the total route and a similar share in traffic. Nevertheless, Bismarck failed in his attempt to create one unified system, the Reichseisenbahnen, though the Länderbahnen were eventually to be brought into such a system in 1920. The shift in ownership is well reflected in Table II.

Technical standardisation of equipment had begun as soon as the separate railways began to fuse together, supervised by the Association of German Railways, formed in 1846 on Prussian initiative. In 1870, there were still, however, 63 railway administrations and over 1,400 different freight tariffs, but a first step to unification came with agreement on a set scale and system of charging by railways in the North German Confederation that year which led to an accelerating pace of agreements on tariffs and through-working throughout Germany. In 1871, the first German timetable conference sought to eliminate long waits at connecting stations and in 1873 a standard time system (Central European Time) replaced local time in railway operation, though the 24-hour clock was not introduced until 1927. 1873 saw the formation of an Imperial Railway Office to supervise legal aspects of railway operation and under its guidance a standard signalling code was defined in 1875 and a major tariff reform begun in 1877. Though through tickets had been available from 1884, it was not until 1907 that the passenger fare structure was unified.

PARTICULARISM AND RAILWAY OPERATIONS

The age of the state railways (Länderbahnen) was imprinted by particularism. An agreement of 1868 between Prussia and Hessen compelled all passenger trains to stop in Offenbach, consequently most fast trains between Frankfurt and Hanau avoided this route. Even until the Second World War, many expresses from Berlin to South Germany continued to run via Halle/Saale instead of via the more important Leipzig, because Prussia had routed its trains this way to avoid crossing Saxon tracks. In 1895 Prussia had brought the Saal-Eisenbahn mainline (opened 1874) and greatly increased the capacity of the Grossheringen-Saalfeld section as part of the Berlin-Halle-Grossheringen-Saalfeld-Lichtenfels-Nürnberg route, with the intent to divert most Berlin-Nürnberg traffic from the Leipzig-Reichenbach/Vogtland-Hof route to spite Saxony. Until 1914 Prussia still preferred to route many Berlin-Köln expresses via Altenbeken-Soest-Paderborn rather than via Hannover, even though the former route was longer and slower and the offending Hannover had been eliminated politically in the war of 1866.

Until 1930 the Berlin-Nürnberg-Ulm-Friedrichshafen expresses ran as semi-fast trains (lower supplementary fare) through Württemberg territory, a condition originally introduced by Bavaria (to reduce competition with its own trains) as the price of allowing the Württemberg railways' line Ulm-Aalen to cross its territory for a few kilometres north of Ulm. Though Prussia had originally opposed building railways along the west bank of the Rhine, it tried hard after 1870 to divert railway traffic along the river on to this bank, so that it would pass over the tracks of the Prussian-controlled Reichseisenbahn in Alsace rather than over the east bank tracks controlled by the Baden State Railway. To counter this, Baden built especially fast locomotives in order to offer the better service.

After 1883, Bavaria had insisted that through tickets for the Orient Express for passengers from Belgium (including England via Ostend) and Luxemburg should be made out not via Strassburg (Strasbourg) in the Reichsland Alsace but via Aschaffenburg-Passau-Vienna, the longest possible route across its own tracks. This was a condition of allowing the Orient Express to run across Bavaria from Stuttgart via Ulm-München-Salzburg to Vienna. In his 1833 pamphlet, List had suggested a direct Erfurt-Nürnberg link. Bavaria had built the southern section from Bamberg to the Thüringen border, while in Sachsen-Coburg-Gotha a line had been built from Coburg south to Rossach near the Bavarian border. Though Coburn became Bavarian in 1920, no attempt was made to build the missing, easily-graded 10 km. line between Rossach and Untermersbach and trains continued to use the heavily graded route via Lichtenfels.

RAILWAYS AFTER THE 1914-1918 WAR

The 1914-1918 War taxed the railways to the maximum and they emerged badly rundown and faced by reparations losses under the peace treaties. Large quantities of rolling stock had to be surrendered, and in several frontier areas (e.g. along the Belgian border), double track had to be singled, while the operating capacity of certain strategic lines reduced. Territorial loss in both east and west reduced the 1914 route length by about 13 percent. The investment loss was considerable; for example, in Alsace and Lorraine, now taken by France, the Germans had trebled the route length between 1871 and 1914; but losses were greatest in the extensive eastern territories incorporated into the new Polish State (Table IV).

The new territorial boundaries created several operating problems, especially in the east where East Prussia was now separated from the main body of Germany by the Polish Corridor. Fortunately, the route of the main Berlin-Königsberg railway via Schneidemühl ran parallel to the frontier for a considerable length, so reducing the distance needed to cross Polish territory. The Poles insisted, however, on onerous conditions of transit, and constant complaints were made about the Polish handling of this traffic, particularly the long delays involved. As relations between the two countries deteriorated after 1935, East Prussian traffic shifted increasingly to a specially instituted shipping service between Stettin and Königsberg (Seedienst Ostpreussen). Elsewhere along the new Polish-German frontier disruption was caused for both German and Polish railways by junctions lying on the opposite side of the new border. The division of the industrial area of Upper Silesia created numerous awkward operating situations in the dense local railway network, but for the Germans, the cutting of the Breslau-Beuthen freight line was a grave inconvenience. With Czechoslovakia, however, several successful working arrangements were made, largely based on agreements

existing with the defunct Habsburg empire. The most notable were
at Asch (Aš) and in the Sudeten districts around Warnsdorf in the
Lusatian Mountains, with German running rights over Czech tracks.

In Schleswig, the return of territory to Denmark left the
Germans without easy access to the island of Sylt until this popu-
lar holiday island was joined to the main German railway system by
the Hindenburg Dam in 1927. A strange anomaly arose after Belgian
annexation of the Eupen and Malmédy districts. The <u>Eifenbahn</u> (Map
12) from Raeren to Kalterherberg and Malmédy was put under Belgian
control (operated by the Belgian state railways) yet it ran through
Germany for much of its length, creating difficult questions of
extra-territoriality. On the new frontier with France disruption
was modest, though until 1935 the Saar railways were outside German
control.

Neither economic nor political conditions favoured further
extensive new buildings, so where new lines were built, the work
had usually been begun or planned before 1914. So far as financial
stringency allowed, improvements were made by double-tracking or by
making working at junctions easier. Nevertheless, several short
but useful freight lines were completed in the Ruhr, notably the
Velbert-Kettwig and Witten-Wengern routes, approved in 1911 and
opened in 1916 and 1927, while in 1934 the long needed Witten-
Bommern-Schwelm route was opened. One vital Ruhr link, the north-
south connection from Wuppertal via Velbert to Bottrop, was never
completed and became one of the first lorry and bus routes instituted
by the railways. Planned in 1913 and opened in 1920, the Dortmund-
Lünen-Münster line provided a useful short-cut from the eastern
Ruhr to the north avoiding the busy junction at Hamm. Though built
for double track, it remained single even after being electrified in
1961. The Celle-Langenhagen route, begun before 1914 but not completed until 1938, provided a time-saving short-cut from the
Hannover area to Hamburg, while completion of the through route
from Bremen via Nienburg to Minden between 1921-1923 was a valuable
additional way from the Hanseatic port. Elsewhere a few short lines
were completed, notably in the Black Forest for tourist purposes,
and in Saxony, where there was rebuilding of narrow gauge route to
standard gauge, though some of the lines were converted more to
relieve unemployment than to extend railway operations.

Late last century, German experiments had shown that electric
traction could be usefully used on railways, but this was opposed
by the General Staff, which felt it made railways too vulnerable to
disruption in war[7]. Consequently, only local routes were electrified,
the first being the München-Starnberg commuters' line in 1882,
followed in 1900 by electrification of commuter routes in and around
Berlin and in 1911, the heavy freight railway between Dessau and
Bitterfeld in Central Germany was electrified, while even during the
First World War there was some electrification of heavily graded
routes in the Silesian uplands. After 1921, electrification again
began piecemeal in Central Germany (e.g. Zerbst-Bitterfeld-Leipzig-
Halle, 1922) and in Bavaria (e.g. München-Garmisch, 1925) and continued in Silesia. By 1939, four local but unconnected systems had

developed (Map 13) - the Breslau-Görlitz system in Silesia; the Berlin radial commuting system; the emerging 'Saxon Ring' freight system in Central Germany; and an extensive South German system radiating from München. A few electrified lines were also running in the Black Forest, as commuting lines in Hamburg, and on the remarkable private Köln-Bonner Eisenbahn, although a plan to create a fast electric commuting railway system in the Lower Rhine-Ruhr area had not been implemented by 1939. The major mainline project to electrify the München to Berlin route was completed from the south to Leipzig in 1942, but Berlin was never reached. Undoubtedly, the availability of hydro-electricity in South Germany and of cheap thermally generated current in Silesia and Central Germany was a key factor in the emergent pattern.

CREATION OF THE DEUTSCHE REICHSBAHN

In 1917-1918, there had been an unsuccessful attempt by the Reichstag to get the states to surrender their railway interests to a unified management, but the old particularism had again triumphed. The Weimar Constitution, Art. 89, called, however, for such unification in one management of all German railways carrying general traffic, in order to strengthen the ties between the regions. In 1920, an agreement between the central government and the eight states with their own railways plus a few remaining large private systems brought the overwhelming proportion of route into a centralised control. The states, now lifeless shells in the Weimar Republic, were quite ready to dispose of their railways, which were generating heavy deficits and no longer serving the intra-state and inter-state political-geographical role of pre-war times. Though the legal and financial organisation of the new national railway system was to change over the years, the so-called Deutsche Reichsbahn (Map 14) made considerable progress in rationalisation and modernisation, especially through the genius of Julius Dorpmüller[8].

With less duplication of routes and other facilities than in Britain, possibilities of rationalisation of the network were limited and the main effort was devoted to locomotives, rolling stock and operating methods. Through services were greatly improved and both passenger and freight traffic accelerated, with the new pattern marked by the introduction of the prestige Rheingold express in 1928, fast diesel railcars on internal routes in 1934, and a massive speed-up of freight traffic in 1935. The new management was remarkably successful in improving the financial fortunes of the railways in the face of growing competition from motor transport[9]. It was perhaps ironical that the new Reichsautobahnen were a subsidiary enterprise of the railways, forced on them by the National Socialists. It was, however, a way of employing the great skill and experience of the railways' civil engineering departments, now underutilised through the ending of new railway construction. In a way, the new motorways were to serve as a complementary arterial transport network to the major railway routes.

Most severely affected in the post-war years were the small railways that had not been taken into the Reichsbahn. The Kleinbahnen, small local enterprises at either standard or narrow gauge, were the first to feel the pinch: by 1935, it was estimated that two-fifths were completely uneconomic. Some 65 percent of their receipts were from freight traffic, as yet little affected by road haulage, but greatly reduced after the 1929-1932 depression. It is not surprising that even several of the more successful remaining private railways were absorbed into the Reichsbahn up to 1943: the less successful were gradually closed or went over to road haulage, though certainly shortage of petrol during the war years helped some to keep operating as steam railways.

THE YEARS AFTER 1945

As territories were incorporated into the Reich from the late 1930's, the Reichsbahn took over the running of their railways - in 1938 the railways of Austria and the Sudetenland were added, in 1939 came the Lithuanian lines in the Memelland and those of western and northern Poland reincorporated into Germany, while in 1940 railways in Eupen and Malmédy and Luxemburg were taken over. At the height of their success in 1942, a further 42,000 km. of operating route in occupied Poland and Russia was run by the special Ostbahn administration. Additionally, the Germans controlled railways elsewhere in occupied Europe and greatly influenced operations on those of their Balkan and Danubian allies. The total length under German administration in one form or another amounted to probably about 250,000 km.

After the capitulation of what remained of German territory, the railways were a mere fraction of this vast system in the brief life of Grossdeutschland. The railways were not only run down, but they were also badly damaged. Annexation by Poland and Russia of Germany's 1937 territory east of the Oder-Neisse Rivers meant a loss of 9,950 km. of route and all related equipment, while the separation from Germany again of the Saar by France reduced the 1957 network by another 550 km. Saar railways until 1937 were run in close association with the French S.N.C.F. Substantial quantities of railway equipment were again to be lost as reparations.

War damage to track was extensive and the cost of restitution was often a critical factor in the decision to abandon lightly-loaded route. The severe limitations applied by the occupying powers on trans-frontier movement brought further closure of railways across Germany's borders, augmenting those closed or abandoned between the wars. On the new eastern frontier, some eighteen routes were affected, though dislocation for the Germans was most

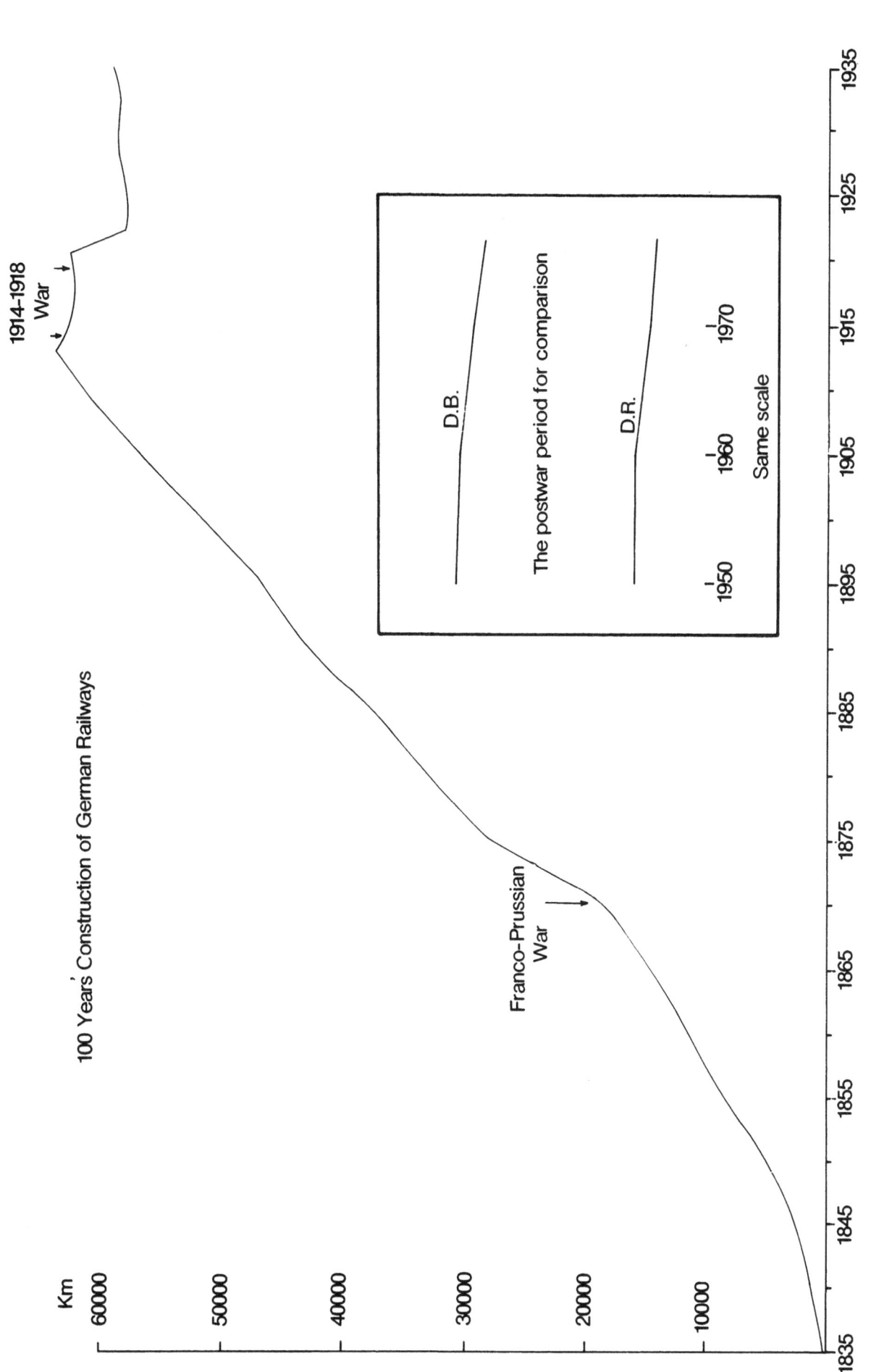

serious on the island of Usedom, whose railways were now cut from
their access to the remainder of the system by Swinemünde (Świnoujscie)
junction passing into Polish hands. At Guben, on the other hand,
the main junction remained on the German side of the Oder; but the
important German freight line between Görlitz and Zittau lay on the
Polish bank of the river and has been operated for the Germans by
the Polish State Railways. On the borders with Czechoslovakia,
numerous local lines across the frontier have been closed or restricted to freight working and the Germans have lost their special
operating rights on Czech territory around Aš (Asch) and Cheb (Eger).
In the west, the return of Eupen and Malmédy to Belgium once again
left the Belgians to operate the Eifelbahn across German territory
until its closure in the early 1950's.

Early Allied discord quickly evaporated the original intent
of treating Germany as one political, economic and social entity.
The occupation zones became self-contained, though the British and
American zones were brought together in the so-called Bizone, which
was later joined with some reluctance by the French, whereas the
Soviet zone became increasingly isolated. Thus at first, the railways in each zone were managed as separate systems, though with such
intense disruption and damage, a high level of local autonomy would
have been necessary in any case. In late 1946, the railways in the
British and American zones were put under a unified management,
while the railways in the French zone were formed into a separate
Southwest German railway administration. In the Soviet zone, the
railways continued to operate under the pre-war title, Deutsche
Reichsbahn, but with the formation of the German Federal Republic in
1949 moves were speedly made towards a unified management of the railways in its territory under the title Deutsche Bundesbahn, legally
formalised on 13 December 1951 (Table V).

The frontier between the two German republics had originally
been intended as a convenient line of demarcation between the western
zones and the Soviet zone: it was never imagined it would become a
true international frontier. Based on pre-war local government
boundaries, it has cut through the landscape with little or no
reference to patterns of economic, social or transport geography;
in fact, it crosses one of the best developed parts of the railway
network before 1945 that carried extremely heavy traffic. As
relations between the two German states have remained strained,
traffic across the new Inter-German border has been light, and the
new states have reorganised and reoriented the direction of traffic
to suit their new spatial patterns. Though some 38 railways are
cut by this border, only nine remain in use, of which two are solely
for freight traffic while only four may be used for traffic between
West Germany and West Berlin (Map 15). Typical of the problems
created is the railway between Eisenach and Gerstungen (both in the
Democratic Republic) which crossed West German territory for several
kilometres, so that the eastern state replaced this section in 1962
by a new line between the two towns via Förtha completely across its
own teritory; in West Germany, Tettau (Kreis Kronach, Bavaria) has
been deprived of its railway branch to the mainline because it ran
for about 15 km. through the German Democratic Republic. Up to

ten railway wagons a day are now carried from the mainline to Tettau on special road vehicles[10].

In the Soviet zone, railways passed under strong Russian influence, relfected in changes in operating procedures. Though damage was generally less severe than in West Germany, the Soviet authorities extracted heavy reparations, taking not only locomotives and rolling stock but also signalling equipment, while several branchlines and lightly loaded cross-country routes, especially in the agricultural areas of Mecklenburg and Brandenburg, were completely dismantled[11]. Double tracked routes, other than those of strategic importance to the Red Army, were singled: in 1950 double or more track route represented only ten percent of the total compared to 37 percent in 1937. After 1950, however, major freight routes were again provided with at least a second track, and by the 1970's extensive replacement of the second track was being undertaken. Attention was devoted in the 1950's to the development of a high capacity strategic route from Thüringen via Karl-Marx-Stadt (Chemnitz) and Dresden to Görlitz; eastwards it continued across southern Poland and the Ukraine and was specifically designed for Warsaw Pact needs[12].

Cut from its main pre-war port of Stettin (Szczecin), now in Poland, and also from Hamburg, now in West Germany, the G.D.R. has had to develop a replacement, for which Rostock was chosen as having the greatest potential. Unfortunately, direct railway links with the industrial south from the Baltic coast were mediocre, so a new trunk line has been under construction, mostly by using existing sections of secondary route linked together to provide through working. The major constructional project has been, however, the completion of the Berlin Outer Ring railway. Begun before 1914, its completion was forbidden under the Versailles Treaty, though work did begin again in 1926 on the Saarmund-Grossbeeren section, and between 1930-1943 the eastern section (Karow-Bohnsdorf-Grossbeeren) was completed. Dismantled in 1945, this eastern section was re-railed in 1948 from Karow to Bohnsdorf, but as the Bohnsdorf-Grossbeeren section now ran through the western sectors of Berlin, a new alignment from Bohnsdorf via Glasow to Grossbeeren, entirely through G.D.R. territory, was completed in 1951. The northern section from Wustermark via Hennigsdorf to Karow was completed between 1950-1959 and a cut-off link to ease movement on the Saarmund-Wildpark section finished in 1957 (Map 16). The line, involving much heavy engineering work, provides a valuable commuting link around Berlin without passing through the western sectors and is also used by G.D.R. long distance trains to avoid West Berlin. Elsewhere other new construction has been limited to short lines to open up the Lusatian lignite field and to a commuting railway for the new residential town of Halle-Neustadt.

Cut from its pre-war supplies of locomotive coal from the Ruhr and from the now Polish Upper Silesian coalfield, the G.D.R. faced a serious fuel problem for its steam railways. Attempts to increase supplies from small home deposits were not outstandingly successful, nor were attempts to use lignite, whose poor calorific value makes steam-raising in an ordinary firebox difficult, while the high ash content creates problems of waste disposal, since ash dropped on the tracks causes bad water-logging. As the electricity supply industry was rebuilt in the 1950's, thoughts turned to railway electrification, but pre-war routes had been dismantled as reparations by the Russians, so work had to start from scratch. Electrification began afresh in 1953, though the München-Berlin route was not again considered, with effort concentrated instead on heavily used freight routes forming the so-called 'Saxon Ring' (Map 17). Within ten years all the principal industrial centres in the Elbe-Saale basin were linked together. It is now proposed to electrify the main freight route via Bad Schandau into Czechoslovakia at Děčín (Tetschen), and in 1976 it was announced that the Leipzig-Berlin route would be electrified. A line over difficult gradients detached from the rest of the electric railways is the Rübelandbahn between Blankenburg and Königshütte in the Harz, used for large shipments of cement, limestone and iron ore. Despite having to import all oil supplies, diesel locomotives and railcars have been increasingly used in place of steam traction. Unlike West Germany which supplies all its own diesel and electric locomotives, the G.D.R. buys in from other Comecon partners.

The G.D.R. railways also operate the railways in West Berlin, except the U-Bahn, whose lines were taken over by the West Berlin authorities when the city was ultimately divided. Even before completion of the Outer Ring railway, the Reichsbahn had, under Russian pressure, begun to divert its own trains away from West Berlin. Only one through route remains in use for international passenger trains between Western Europe, Poland and the U.S.S.R. One or two routes (varying over time) have been available to through trains from West Germany to West Berlin. As a result of these changes several pre-war long distance stations have been closed and some secondary routes in and around the city dismantled (Map 16).

The larger part of the railway network remaining in Germany after 1945 passed to the German Federal Republic's Deutsche Bundesbahn (Table VI). This system contains the busiest of the pre-war routes on which further substantial increase in traffic took place after 1945: in the G.D.R., in contrast, post-1945 traffic has only recently reached let alone exceeded pre-war levels. The development of the West German economy and its integration into the broader framework of the European Economic Community resulted in a reorientation of traffic onto a predominantly north-south axis compared to the east-west alignment of pre-war times. Though West German railways suffered greater war-time damage than those in the G.D.R., they lost less equipment through reparations. In the French zone, however, the authorities tried to divert all north-south international traffic along the French bank of the Rhine via Strasbourg rather than along the German bank via Karlsruhe and Freiburg/Breisgau by reducing the mainline in Baden to single track at certain points. Continuing German obligations in international traffic (notably that passing through the Swiss alpine tunnels) brought pressure from the British and Americans that forced the French to abandon their policy.

Badly damaged lines with a record of only light traffic were
generally not rebuilt, while by the early 1950's little used but
operational rural railways (notably narrow and standard gauge
Kleinbahnen) were being closed. Some purely strategic bridges
(e.g. on the Rhine at Remagen and Bingen) or others with light
traffic (e.g. Wesel) destroyed in the war were not rebuilt. Few
truly new lines have been built - the most important has been from
Buer-Nord via the new town of Marl to Haltern, giving better access
from the northern edge of the Ruhr coalfield to North Germany. The
Hannover-Langenhagen-Celle line has been double-tracked to carry
growing north-south traffic in the Hannover area, while from
Braunschweig and Wolfsburg the Wittringen-Wieren-Uelzen route has
been greatly improved to give better access to Hamburg. The need
to speed traffic to Scandinavia by eliminating the long, slow route
via Flensburg and Fredericia to Copenhagen was solved by providing
direct train ferry access from Puttgarden on Fehmarn to Rødby Havn
on Lolland, including a long road-rail bridge over the Fehmarn Sound
(1963). This so-called Vogelfluglinie replaced an interim train
ferry (1951-1963) from Grossenbrode to Gedser that was introduced as
the pre-war route via Warnemünde was no longer accessible because it
now lay in the G.D.R.

Much of the new investment has been in commuting lines in the
major towns and conurbations. The long postponed fast electric
commuting system for the Ruhr discussed in the 1920's has been started,
with trackwork for the new regular interval S-Bahn trains separated
from that for freight and long-distance passenger traffic. The new
commuting lines have in several instances involved short lengths of
new route in costly urban situations, like the east-west line in
München and the Chorweiler branch in Köln or the Nürnberg system,
while in Frankfurt am Main a special line to serve the Rhine-Main
airport has been built. Extensive rationalisation of trackwork at
junctions has been undertaken and major station reconstruction
carried out. In 1955, reconstruction at Heidelberg station
eliminated the serious traffic obstacle of dangerous level crossings;
shortly after, a new and enlarged station was opened at Bochum;
while the operationally difficult terminal station at Braunschweig was
rebuilt as a through station; and an interesting new layout for
through running was completed at Ludwigshafen in the mid-1960's.

Strategic interest in the railways has, however, not been lost.
The operationally costly but modestly used Wutachtalbahn around the
Swiss salient at Schaffhausen has been closed to through civilian
traffic, but it was for many years maintained in operational readiness with NATO support. The same condition of readiness also seems
to apply to secondary routes along the eastern border with the G.D.R.
and Czechoslovakia, though it is uncertain whether NATO has also
supported these lines.

The considerable realignment of traffic on to the north-south
axis resulted in growing pressure on existing route facilities.
While the combination of fast accelerating diesel and electric
traction, computerised timetabling and more sophisticated signalling
equipment made possible big increases in carrying capacity on exist-

ing routes, there have nevertheless been limits; as an example, on
the Rhine section between Köln and Koblenz, these new methods made
it unnecessary to quadruple track as planned in the late 1930's,
and traffic has been increased from 74 trains a day in 1938 to 199
in 1952 and to 326 trains daily in 1976. By 1970 it had, however,
become apparent that better north-south links were needed and a plan
was outlined to overcome the inadequacies and bottlenecks in the
existing system in relation to expected speeds and traffic density
by the 1990's. Wherever possible existing tracks would be used,
but some new sections were nevertheless to be required. Several
variants to the scheme were proposed and some elements in the plan
made conditional on long term developments in traffic. The first
stage, to be completed by 1985, will be followed by a second phase
with an unspecified completion date. The completely new routes
and those to be developed from existing lines are shown in Map 18:
this involves considerable sections of new track between Hannover
and Würzburg (on which some work has already begun) and a completely
new railway over the hills on the right bank of the Rhine between
Köln and Frankfurt. Work on the direct Mannheim-Stuttgart link is
already underway and rectifies a long-standing deficiency from the
days of the Länderbahnen. The second phase, if realised, could in-
clude a modern version of the missing Weser-Bahn from Bremen via
Löhne, Bielefeld, Paderborn and Marburg to Frankfurt as well as two
east-west links (Kassel-Soest-Düsseldorf and Nürnberg-Crailsheim-
Stuttgart or Mannheim) to eliminate present circuitous routes.
The readiness to make the necessary massive investment in new and
improved routes suggests a strong belief in the long term role of
the railways as a medium and long distance haulier.

In 1949 plans were formulated to extend the pre-war system of
electrified railways and a steady progression was made thereafter
from South Germany. By the early 1960's, these had spread down
the Rhine to the Ruhr area and in the second half of the 1960's
the links north to Hamburg were completed. By 1978, with plans
nearly completed (Map 19), well over 10,250 km. of the total 29,000
route kilometres of the network were using electric traction,
handling over 80 percent of the total traffic. Fast post-war
express services were provided initially by diesel railcar sets
(including the abortive Talgo-type Gliedertriebzug) and these were
used to develop the original Trans-Europ-Express (TEE) network.
Fluctuations in traffic and maintenance problems led, however, to
their replacement by light locomotive-hauled trains (initially
steam, but later diesel and electric traction), but interest in the
mid-1970's has swung back to the railcar-type trains similar to the
British Advanced Passenger Train (APT). In 1977 steam traction was
finally abolished after a steady retreat onto the coalfields, and
diesel traction is used on all non-electric sections. On lightly
loaded rural routes, battery railcars have been used with some
success, but the interchangeable road-rail vehicle proved a failure.

Financial losses on the railways mounted through the 1960's,
exceeding those on British Railways. The loss arose from the in-
roads of other hauliers (notably the private car), but there was
also the rising unit cost of labour and the continuation of outdated

manning practices. Another cause was the railways' obligation to
continue loss-making operations, notably the heavy losses arising
from unnecessarily large fare concessions for many types of commuting traffic. Additionally, there was the high cost of renovation
and repair of the war-damaged system, to which also had to be added
the cost of modernisation, despite help from the federal and Länder
governments. Although there was the growing challenge from road
haulage and the late but telling emergence of competition from
airways, the German government - at federal and Land level - maintained a remarkable faith in the long term prospects for the railways. Growth in many traffics was supported by optimistic long
range forecasts. Intensified economic activity brought a rise in
freight traffic, while growing affluence saw more medium and long
distance travellers, despite greater car ownership. By the time
of the energy crisis in 1973, forecasts suggested that freight
traffic would grow by 50 percent and passenger traffic by 100 percent by 1985.

Retention of passengers and even some increase has been encouraged by good punctuality, clean and efficient service, besides
a conveniently timed service from early morning to late at night
supported by an attractive fare structure to cover most types of
traveller. As noted, the black spot has been commuting traffic.
Freight traffic growth has been helped by modern handling and
rolling stock plus guaranteed and express delivery services.
Containerisation and piggyback trains have also been of increasing
importance, while government assistance has come in the form of
grants for the building of private user sidings and the prohibition
of lorries using motorways over weekends.

Nevertheless, by the mid-1970's the need for a rigorous
rationalisation of railway operations was apparent. Though since
1945 some 5,200 km. of route has been closed, overwhelmingly on
lightly used rural lines, a further reduction of 6,000 km. of route
for passengers and some 3,000 km. for freight traffic is now planned[13].
In this process, some mainlines (Hauptbahnen) will be reduced in
status to secondary lines (Nebenbahnen), with consequent savings in
maintenance and equipment costs. Uneconomic stations will be
closed and the number handling parcels traffic reduced. In many
instances, buses will replace lightly used passenger trains. Indeed, the wide network of buses operated by the Bundesbahn and
Bundespost has contained the threat from competing bus and coach
services. Some of the non-federal railways (e.g. the Westfälische
Landeseisenbahn) have already shown how simplified but rationalised
operations can ensure survival. On the other hand, in order to
cater for the expected increases in traffic on arterial routes, a
major construction programme for entirely new high-speed routes and
the massive improvement of existing routes has been instituted.
Mainline services are being tailored to suit modern requirements
and expectations - fast, regular interval service with a high level
of comfort.

The Deutsche Bundesbahn remains one of the most efficient and
effective systems in Europe. With the overwhelming bulk of traffic

hauled by electric traction, it is being metamorphosed into a railway for the twentyfirst century. This has been helped by a consistently realistic policy towards railways, without the vacillations and uncertainties faced by British Railways under successive governments. Though the Bundesbahn has had no Beeching Report, many German railwaymen accept that the principles proposed by Beeching are equally applicable to their system. The Bundesbahn has, however, been in a possibly easier situation, because there are fewer old declining areas of heavy industry with over-extensive and outdated local networks and little legacy of duplication left by the old companies; there is also in the Federal Republic a more even spatial distribution of traffic generation without the dominating pull exerted by London; moreover, distance terms make railway traffic highly attractive in many international movements in Europe, so attracting traffic from within Germany or from outside across its tracks. On the other hand, the Bundesbahn has had to face the conflicting interests of the Länder. Consequently questions of rerouting or line closure can be toughly fought by Land governments if they feel they are being less favourably treated than their neighbours; or they can force electrification or other investment in routes not considered worthy on purely railway operational grounds.

ASPECTS OF TRAFFIC AND SERVICES

Predominantly single track and simple signalling arrangements limited the volume of traffic handled by the early railways, when two or three trains a day seem to have been the general level. An early timetable of the Berlin-Hamburg Railway shows two through passenger trains and one through goods train. An afternoon train of type unspecified ran as far as Wittenberge and continued next morning to its destination, largely because of a reluctance to run in the dark. On the Dortmund-Elberfeld line in 1849 there was only one through train and four others covering part of the route. Travel times were relatively slow: Berlin-Frankfurt/Oder took 3.5 hours, Berlin to Stettin or Magdeburg 4 hours, Berlin-Hamburg 8 hours and Berlin-Breslau 13 hours. The first night passenger train ran from Berlin to Köln in 1848 but needed 24 hours for the journey: this had been reduced to 9 hours by 1896 and eventually to 5 hours in the 1930's. The first advertised express train (1851) was also from Berlin to Köln, covering the distance in 16 hours and providing a through connection to Paris. In 1852 light 'courier trains' were introduced, reducing the Berlin-Köln time to around 14 hours, and by the 1880's a comprehensive network of these 'courier trains' (Map 20) served all the principal places.

By 1855 nineteen of the 28 railway companies in Prussia had advertised express trains, when one-ninth of all timetabled trains were in this class, one-third were stopping trains, one-third goods trains and the remainder, mixed trains.

The sleeping compartment was introduced in 1851 and the first proper sleeping car ran in the Berlin-Ostend express in 1873, provided by the International Sleeping Car Company, which later ran a Köln-Paris service. On internal services, this company operated only in South Germany: Prussia, Mecklenburg and Saxony and the railways in Alsace and Lorraine provided their own cars. After 1885 Prussia began to restrict the company's operations wherever a Prussian car could be substituted. In 1880 the first restaurant car service ran between Weimar and Eisenach, again operated by the International Sleeping Car Company. The service was later extended to Berlin and Bebra, but an attempt to extend it to Frankfurt am Main was vetoed. Restaurant cars were soon being run by German railways, notably by Prussia and Baden and by two special companies (German Dining Car Company and Northwest German Dining Car Company) as well as on a local scale by three other companies formed by former station restaurant owners whose trade had been undercut. By 1914 the International Sleeping Car Company was running some 28 international and 39 internal services across Germany, but in 1916 these along with other sleeping and restaurant car companies were taken over by Mitropa (Central European Sleeping and Dining Car Company): in 1925 Mitropa, curtailed in its operations by the peace treaties, made a traffic-sharing agreement with the International Sleeping Car Company.

By the late 1840's agreements between companies to standardise equipment to make through running possible were being concluded and in 1848 a timetable for all German railways appeared. The main growth in traffic came, however, as industrialisation grew in the 1860's. By 1877 it was claimed that courier trains covered 8,684 km. of route, expresses with only first and second class accommodation ran on 2,765 km. of route and fast trains with three classes of coaches covered 4,857 km. of route. Out of a total route length of 30,452 km. at this time, some 14,146 km. had only ordinary stopping trains (which on some railways conveyed even a fourth class). Traffic rose sharply in the last two decades of the century and continued a healthy upward trend until 1914, when an effective network of express and semi-fast trains was operating (Map 21). The Orient-Express in 1883 had set a new style in international trains so that by 1914 a series of luxury trains linked Berlin to other capitals as well as to high class spas. One of the most important, introduced in 1896, was the Nord-Express from Paris via Köln, Berlin and Eydtkuhnen to St. Petersburg and Moscow. Speeds were usually modest and some railways (like Prussia) preferred punctuality (Table VIIa & VIIb). The fastest trains ran at up to 90 k.p.h., mostly in the flat terrain of North Germany (notably Berlin-Hannover, Berlin-Hamburg and Berlin-Halle). Baden, with a long well graded mainline along the Rhine, was also noted for its speedy trains, but the rougher terrain of Saxony, Bavaria and Württemberg put them in a lower speed table though some notable services did operate. The services on secondary routes in both speed and frequency had by 1914 reached a level that was to remain little changed until the 1950's.

The 1930's witnessed the speeding up of most mainline trains, but the most striking acceleration came with the use of fast diesel railcar sets, of which the Fliegende Hamburger of 1934 was the forerunner. The reasons underlying the new accelerations were competition from Germany's well developed air services; the general search in Europe and America at the time for railway speed records; and in the case of the railcar services (some observers claim), a desire by the National Socialist government to have a tighter hold on the provinces, since most of these services radiated from Berlin (Map 22). This was also the zenith of the through coach services. Where a full train was not warranted, through coaches to a particular destination were attached to two or more trains along the course of their route. This type of running demanded good timekeeping and quick marshalling of trains at intermediate stations - examples are shown in Map 23. Since 1945 this type of operation has been increasingly replaced by guaranteed connections, though it is far from eliminated entirely.

Passenger traffic since 1945 has had to be considerably reorganised as a result of the division of Germany. These changes are clearly reflected in Map 24, where the fall in the density of passenger trains in the G.D.R. is particularly evident. In the Federal Republic, the figure illustrates the considerable realignment on to a north-south axis, with the Rhine route standing out forcefully. Whereas after the 1914-1918 War attempts had been made to re-route international trains around Germany (for example, with the long circuitous route chosen for the Orient-Express), no such clear policy was followed after 1945. The Paris-Copenhagen and Paris-Vienna expresses were quickly reinstated and by 1949 a network of international trains again crossed Germany. Services across the 'Iron Curtain' were, however, sparse and liable to sudden withdrawal. Several famous pre-war trains like the Karlsbad-Express and the Paris-Berlin-Riga-Express had not unexpectedly disappeared. The Skandinavien-Express followed the preference of travellers for the long journey via Flensburg and Fredericia to the shorter pre-war routes that would have now taken them across Soviet-held territory, until the new Vogelfluglinie in 1963 greatly improved this journey. The reinstitution in 1951 of the famous Rheingold-Express (first run in 1928) marked a new phase in international traffic, whereas in the G.D.R. the increase in services with other Socialist bloc countries at this time marked the new orientation of traffic, typified by the elaborate routing of the Balt-Orient-Express (Berlin-Bucharest) or the Pannonia-Express (Berlin-Sofia).

In West Germany, two clear patterns of long distance traffic have emerged - the heavy and modestly fast expresses on internal services (Lindau-Kiel, Dortmund-München) or international routes (Austria-Express, Ost-West Express), and the system of high-speed Trans-European Expresses (TEE) started in 1957 to link the western capitals and main centres together (Saphir, Mediolanum) or the more recently instituted accelerated internal Inter-City services. In the 1950's and 1960's there were already a series of light long distance named trains classified as F-Züge on which additional supplementary charges were made, but in 1971 this system was replaced by fast Inter-City trains on a nearly regular interval basis, joining all the main towns (Map 25), while in 1973 a system of connecting

trains (DC-Züge) to the Inter-City network was started. Since the
mid-1970's the daytime TEE-trains have been matched by comparable
sleeping car trains (TEEN). The best trains now all attain speeds
of 100 k.p.h. or more and 200-250 k.p.h. services are planned as
suitable track is extended. Map 26 reflects the centrality of
West Germany in the European railway system, and shows how it also
plays a central role in U.I.C. plans for the future development of
railway traffic based on the principle of direct competition with
air and fast motorway journeys[14]. Timetables are to be based on
timings offering better than two-thirds of the motorway time
(average speed 90 k.p.h.) or city centre-to-city centre time for
air journeys.

The backbone of services remains the semi-fast Eilzüge and
the faster, long distance expresses (D-Züge). Some Eilzüge cover
considerable distances, e.g. from Trier to Hamburg or from Frankfurt
to Passau. Daily, some 450 Eilzug and over 500 express services
are offered (Map 27). The pre-war elaborate system of zone-
related supplements for the use of these trains has been dropped,
partly because the pre-war long distance slow trains (e.g. Frankfurt-
Köln, Düsseldorf-Hannover, Hamm-Norddeich) are no longer run and
partly to encourage rail travel. Instead, supplements are now
charged on these trains for journeys below 50 km. to keep them free
of short distance travellers. The number of ordinary stopping
trains has decreased, especially on the mainlines, and the pre-war
runs of 200-250 km. stopping at every station eliminated. In the
major conurbations, systems of Nahschnellverkehr using fast, light
trains (stopping for less than a minute at most stations) have been
introduced.

In the G.D.R., it was only during the 1960's that passenger
services began to rise to pre-war levels and to overtake them on
selected routes, notably those from East Berlin to the southern
industrial towns. Speeds have remained modest and operating patterns
similar to pre-war, though like West Germany, the stopping passenger
train has been replaced increasingly by the Eilzug[15]. In 1976, a
plan was announced to provide a system equivalent to the West German
Inter-City trains, but like other East bloc countries, passenger
traffic until recently received a lower priority than freight. The
equivalent to the West German TEE services are fast diesel railcar
sets (mostly of Hungarian construction) from East Berlin to the
capitals of other Comecon countries. There is no similar service
to the motorail-type trains (Autoreisezüge) run in West Germany, but
chartered tourist trains like the West German Touropa and Scharnow
trains run to destinations mostly on the Black Sea coast or Lake
Balaton. Services between the two German states are sparse (Map 28),
usually with only one or two trains a day.

FREIGHT TRAFFIC

Because railway construction began before industrialisation got under way on a large scale, freight traffic was slow to develop. All the early railways were designed to depend primarily on passenger traffic for revenue. The first freight on the Nürnberg-Fürth railway was moved six months after opening - appropriately two barrels of beer! By 1848, however, there was a through freight service from Köln to Leipzig, but it needed four days as the trains stopped overnight in Dortmund, Minden and Braunschweig. By 1860 the North German Railway Union had organised through coal trains from the Ruhr to Magdeburg, Berlin and the Netherlands on a special tariff basis. In 1868, the Association of German Railways agreed on a code for the mutual use of wagons, while in 1877 a 'reformed tariff system' was introduced to cover the growing length and range of freight movements. In general, freight rates were structured on a distance basis with fixed level terminal charges on all distances over 100 km. Nevertheless, special tariffs grew more common and in 1909, for example, 61 percent of all freight moved under such Ausnahmetarife. In contrast to Britain where the private owner's wagon inhibited progress with such things as continuous braking and larger capacity wagons, the German railways themselves had virtually complete control of the wagon fleet, so being able to press ahead with improvements. South German railways were particular innovators, with the early use of bogie goods wagons, simple forms of transferable containers, the movement of loaded horse-carts by rail, and even refrigerator wagons using blocks of ice. The rapid upsurge of the chemicals industry in the 1880's brought the growth of a fleet of tank wagons.

The first gravity marshalling yards were developed in Saxony - Dresden - Neustadt 1846, Leipzig 1851 and Zwickau 1862. In 1872 the Rhenish Railway built the large Köln-Gereon yard, while at this time the first true hump yards were built in the Ruhr at Speldorf and Fintrop, with the large Osterfeld-Süd yard opened in 1889. By 1912 marshalling facilities in the Ruhr were becoming overloaded and to cope with the rapidly rising traffic, the inner yards near the mines concentrated on local sorting, while long distance traffic was sorted in peripheral yards, like Duisburg-Wedau and, the largest of all, Hamm, opened in 1916. The years before 1914 also saw the building of strategically placed yards along the east bank of the Rhine, a factor in the concentration of freight traffic on that bank.

Figures for the Railway Directorate Essen reflect freight operations in 1913. About 29,000 13-ton open wagons were available daily, 83 percent used for coal shipments. Only about a fifth of the coal shipped was retained in the directorate: the figure today is much higher. About 1,000 trains a day were dispatched, whereas today the figure is about 400, but it should be recalled that gross train weight in 1913 was generally below 1,075 tons compared to 1,800-2,200 tons nowadays. Whereas in 1913 over 80 percent of the coal shipped from the Ruhr went by rail, the comparable figure today is under 50 percent, reflecting the bigger participation of inland waterways.

After the first war, freight traffic took several years to reach the level of 1913. The change was generally, however, unevenly spread, as reflected by the statistics for various traffic districts in Table VIII. While the total tonnage generally remained below the 1913 level, a change in the structure of shipments and some modifications in the directions of the flows combined to increase traffic (measured in ton-kilometres) in the post-war years. Most notable was the decline in tonnage moved in heavy industrial districts like the Ruhr in the west and Upper Silesia in the east, but substantial increase in tonnage came in the lignite mining areas of Central Germany. The slow recovery of inland waterway traffic, besides coastal and overseas shipping, helped the railways maintain their tonnage, but a sudden serious decline came with the depression in 1929-1932. Even by 1936, a year marked by neither boom or slump conditions, railway tonnage was still only 96 percent of the 1913 level.

During the interwar years, two-thirds of the tonnage handled remained in solid fuels and ores, with mineral building materials and rising volumes of petroleum goods forming most of the remainder. Sugar beet and potatoes comprised the bulk of the foodstuffs shipped, but were marked by strong seasonal fluctuations in volume and, for sugar beet, relatively short hauls. By the outbreak of war in 1939, over 500 million tons of goods were being moved annually, about twice the volume handled in France. At this time, over 90 percent of all tonnage moved in internal shipments, a proportion that had risen during the 1930's as economic nationalism, autarkic policies and the generally depressed state of international trade had intensified. The reviving economy of the late 1930's was marked by rising shipments of coal and lignite, though some of the increase was taken from the railways by inland waterways. The pursuance of self-sufficiency policies increased shipments of home ore, notably in North Germany and the Central Uplands. The building of the motorways and other big projects also boosted shipments of mineral building materials by rail.

Map 29 from the important volume <u>Hundert Jahre Deutsche Eisenbahnen</u> illustrates the significance of the major industrial districts in freight traffic. Indeed, six traffic districts, all industrial areas, carried 40 percent of the total traffic, while almost a fifth came from the Ruhr alone. The proportion held by the Ruhr of traffic moving within one district and not between districts was considerably higher, amounting to almost a third of the national total - in the Duisburg traffic district, for example, most coal shipments moved only a few kilometres from the mines to the Rhine harbour at Duisburg-Ruhrort. Even at this time (1935-1938), unit trains composed of special wagons were being used, as for example, movement of coal in 60-ton bogie hopper wagons forming 1,700-ton trains between Upper Silesia and the Berlin-Klingenberg power station. During the 1930's, with economic nationalism well entrenched, official policy encouraged the introduction of special tariffs to deflect shipments to German ports rather than to foreign ports closer at hand; for example, shipments from the Ruhr and Lower Rhine were cheaper sent to Bremen, Emden or Hamburg than to Rotterdam or Antwerp.

Freight traffic in the 1930's was particularly well analysed by
O'Dell in his contribution on railways to the Naval Intelligence
Geographical Handbook, <u>Germany</u>, Vol. IV, London 1945.

About 1935 a freight modernisation programme was introduced.
Train loads were increased and time spent in marshalling yards reduced, while the better design of wagons and installation of continuous braking made higher train speeds possible. Rationalisation
also played a part; for example, 35 marshalling yards dating from
the old company days in the Essen directorate were reduced to five,
including the modern yard at Hamm (then one of Europe's largest).
Acceleration of goods trains took place across the <u>Reichsbahn</u>:
24 hours were cut from the Passau-Friedrichshafen journey, 15 hours
from Warnemünde-München, and some 14 hours between Köln-Gremberg and
Hof. Effort was made to clear special routes for freight - as far
as possible, freight was concentrated on the east bank of the Rhine
and passenger trains on the west; the Oebisfelde route from Hannover
to Berlin was kept primarily for freight; while special freight
lines were designated round crowded junctions and trackwork improved
to avoid conflicting movements. Some of the major freight routes,
while designed to be the cheapest operationally, were not always the
shortest distance routes and were not uncommonly apparently
circuitous.

The division of Germany has meant the realignment of freight
flows, just as it has for passengers. New flows have emerged as
the new spatial patterns of the two economies have developed. In
West Germany, the shift from coal to petroleum has been marked in
falling shipments of the former, but rising shipments of the latter
have tailed off as the pipeline system for crude and finished products has been extended. In the Federal Republic, railway freight
followed the upsurge in the economy after the currency reform of
1948, but it had begun to level off by the mid-1950's as competition
from road haulage sharpened. To compete, new methods of handling
traffic have been developed, of which containerisation has been most
important, though the early railway designed containers have given
way to international standard modules. During the 1970's, container traffic, modelled on the British <u>Freightliner</u> system, has
grown at about 20 percent per annum (Map 30). Much of the traffic
is between inland centres and ports, both German and Dutch. Most
recently, considerable numbers of containers have been arriving
from Japan via Siberia and the G.D.R., while in 1977 a service from
Bochum-Langendreer to Persia via the Lake Van route was begun.

Following American experience, a successful system of 'piggyback' transport of road vehicles has been developed. This is
easier in Germany than Britain because of the larger loading-gauge.
With growing road congestion and restrictions on the use of motorways by heavy vehicles over weekends, this system has many attractions,
as well as fitting the European Community's policy for the integration
of transport media. Some 17 piggy-back stations over the network
offer 29 nightly services, including a Köln-Brenner-Verona link.
There has also been the increased use of unit-trains: among these
are 4,000-ton ore trains from Emden to the Saar and from Nordenham

and Hamburg to Salzgitter, while 3,000-ton coal trains run from the Ruhr to power stations in Franconia. Consignments of motor cars from inland producers to export ports are similarly handled. An interesting development has been shipment of molten metal by rail from the Rheinhausen steelworks to the Bochum steelworks, some 50 km. away. Though postal railway traffic developed early, the whole system has been speeded up and in 1977, to compete with airways, a network of twenty special postal freight trains each night has been run at speeds over 120 k.p.h.

International traffic, particularly with partners in the Economic Community, has grown, but West Germany has become a focus of the new TEEM network, the freight equivalent of the TEE trains. Over 130 TEEM trains run and of a total 123,961 km. of route operated, 51,245 km. are in the Federal Republic. Of the 91 trains running through the Republic, some 14 operate over through routes exceeding 2,000 km. The longest single route is from Hälsingborg in Sweden across Germany to Sagunto in Spain, some 2,784 km. Köln is the main TEEM junction. In the G.D.R., TEEM traffic is gathered and marshalled at Seddin, southwest of Berlin, where other international traffic is also sorted. Much of the TEEM traffic from Sweden passes through the G.D.R. via the Sassnitz train ferry. Otherwise international ferry traffic is mostly with Comecon countries and the volume with Czechoslovakia is particularly substantial via Bad Schandau-Děčín.

Whereas in the Federal Republic the reorientation of freight traffic has been from a predominantly east-west to a primarily north-south axis (Map 31), the pattern in the G.D.R. has been on to an inner ring of movement in the Elbe-Saale basin, the so-called Saxon Ring (Map 32). The loss of Stettin (Szczecin) to Poland and the cutting off of Hamburg by the Inter-German frontier has also increased freight shipments to the Baltic ports of Wismar and particularly Rostock. The directorates of Halle and Cottbus dispatch about 60 percent of the total tonnage, but receive only about 30 percent. This arises from the large shipments of lignite from these districts, but there are also big shipments of salts from Halle and of mineral building materials from Cottbus. The directorates of Dresden, Halle and Merseburg receive about 60 percent of all tonnage, though no particular items predominate. The Dresden directorate unloads about 30 percent of all wagons daily and is the focal point of distribution of empty wagons to the rest of the system. There has been a growth of container traffic, though not yet on the scale of West Germany, with most traffic between the Saxon industrial towns and Rostock port.

Although the railways of the two German states appear to be growing apart in several aspects (the most obvious direction in their network patterns in relation to traffic flow); on the other hand, the pattern of technological development has been remarkably similar. Nevertheless, the West German Bundesbahn gives its priority to passenger traffic, attempting to run more of its freight services overnight, whereas the East German Reichsbahn gives freight traffic the preference. Perhaps by nature of operating in a planned economy with fixed targets, the latter has been less responsive to change in demand than the Bundesbahn, more open to challenge in a market economy.

TABLE I

THE MOVE TO STATE RAILWAYS - THE PRUSSIAN EXAMPLE

Year	Railway	Length	Notes
1850	Westphalian Rly.	595 km.	} original state-owned lines
	Saarbrücken Rly.	267 km.	
1851	Prussian Eastern Rly.	2,558 km.	
1852	Lower Silesian-Mark Rly.	760 km.	
1866	Hannover State Rly.	1,041 km.	} gained under Treaty of Prague
	Nassau Rly.	189 km.	
1876	Halle-Kassel Rly.	} 224 km.	
	Halle-Nordhausen Rly.		
1877	Berlin North Rly.	228 km.	
1878	Frankfurt-Bebra Rly.	281 km.	(also known as Electorate of Hessen State Rly.)
1879	Berlin-Stettin Rly.	961 km.	
1880	Köln-Minden Rly.	1,108 km.	} total 2,664 km.
	Berlin-Potsdam-Magdeburg Rly.	260 km.	
	Rhenish Rly.	1,296 km.	
1882	Berg-Mark Rly.	1,336 km.	} total 2,272 km.
	Berlin-Anhalt Rly.	431 km.	
	Berlin City Rly.	12 km.	
	Rhine-Nahe Rly.	121 km.	
	Berlin-Görlitz Rly.	218 km.	
	Cottbus-Grossenhain Rly.	154 km.	
1883	Mark-Posen Rly.	282 km.	} total 1,068 km.
	Breslau-Schweidnitz-Freiburg Rly.	600 km.	
	Bremen Rly.	186 km.	
1884	Tilsit-Insterburg Rly.	54 km.	} total 1,064 km.
	Oder Right Bank Rly.	336 km.	
	Posen-Kreuzburg Rly.	201 km.	
	Berlin-Hamburg Rly.	450 km.	
	Kreis Oldenburg Rly.	23 km.	
1885	Braunschweig Rly.	357 km.	} total 659 km.
	Halle-Sorau-Guben Rly.	302 km.	

TABLE I (cont.)

Year	Railway	Length	Total
1886	Magdeburg-Halberstadt Rly.	1,026 km.	
	Oels-Gnesen Rly.	161 km.	
	Thüringen Rly.	504 km.	total 3,445 km.
	Altona-Kiel Rly.	299 km.	
	Upper Silesian Rly.	1,455 km.	
1887	Berlin-Dresden Rly.	181 km.	
	Aachen-Jülich Rly.	40 km.	total 343 km.
	Nordhausen-Erfurt Rly.	122 km.	
1889	Friedrichsroda Rly.	9 km.	
1890	Werneshausen-Schmalkalden Rly.	7 km.	
	Schleswig-Holstein Marsh Rly.	238 km.	total 448 km.
	West Holstein Rly.	100 km.	
	Lower Elbe Rly.	103 km.	
1895	Weimar-Gera Rly.	69 km.	
	Saale Rly.	94 km.	total 375 km.
	Werra Rly.	212 km.	
1896	Hessen Rly.	199 km.	Also known as
	Hessen Ludwigs Rly.	741 km.	Main-Weser Rly.
	Hessen State Rly.	221 km.	total 1,161 km.
	(Creation of the Prussian-Hessen State Rlys.)		
1902	Eisenberg-Crossen Rly.	9 km.	total 101 km.
	Main-Neckar Rly.	92 km.	
1903	Dortmund-Gronau-Enschede Rly.	103 km.	
	East Prussian South Rly.	246 km.	
	Marienburg-Mlawka Rly.	149 km.	total 701 km.
	Kiel-Flensburg Rly.	81 km.	
	Altdamm-Colberg Rly.	122 km.	
1904	Breslau-Warsaw Rly.	56 km.	
1914	Cronberg Rly.	10 km.	
1918	Prussian Military Rly.	70 km.	

Source: Various.

TABLE II

HISTORICAL STATISTICS OF GERMAN RAILWAYS

	1870	1890	1913	1925	1930	1936
Population (mill.)	41.26	49.24	66.98	63.1	65.1	66.9
Railway length (km.)	18,805	41,495	63,794	57,268	58,344	58,991
Of which state railways	8,627	38,105	58,976	52,768	53,844	54,491
State administered private railways	1,969	98	106	–	–	–
Private railways	8,210	3,292	4,712	4,500	4,500	4,500
Passengers (mill.)	113	421	1,834	2,106	1,829	1,611
Tonnage (mill.)	70	213	559	409	400	452

Source: Various.

TABLE III

RAILWAY OPERATIONS IN PRINCIPAL GERMAN STATES 1885 and 1910

	Prussia-Hessen	Reichsland, Alsace & Lorraine	Bavaria	Württemberg	Baden	Saxony	Oldenburg	Mecklenburg
Operating length (km.)	21,050[1]	1,490	4,350[2]	1,536	1,318	2,078[4]	370	349
	37,349	2,007	7,669	1,930	1,750	2,839	663	1,100
Population ('000)	28,314	1,563	5,416[3]	1,995	1,601	3,179	341	575
	45,437	1,871	6,876	2,436	2,142	5,019	482	640
No. of trains ('000)	2,671.7	358.5	231.0	123.3	172.9	371.0	27.3	20.9
	9,902.5	980.8	1,579.7	501.2	812.9	1,154.1	128.6	113.9
Train-km. (passengers) (mill.)	86.9	6.5	17.4	5.9	6.6	9.8	1.0	0.9
	290.9	18.4	47.8	16.7	17.0	24.7	3.2	4.7
Train-km. (freight) (mill.)	65.7	5.2	3.7	1.9	2.7	6.5	0.4	0.3
	180.4	12.1	22.7	6.4	9.2	11.7	1.5	1.7
No. of passengers (mill.)	161.9	12.2	18.4	12.2	13.0	22.6	2.3	1.3
	1,083.8	48.1	121.4	64.7	53.0	103.6	9.4	7.2
Passenger-km. (mill.)	5,030.3	301.2	650.4	280.7	304.2	583.8	54.1	59.8
	25,221.9	1,201.4	3,273.7	1,202.3	1,174.9	2,186.3	221.2	270.8
Freight tonnage (mill.)	97.6	9.7	8.5	3.4	5.6	12.1	0.7	0.05
	390.5	40.1	40.2	13.2	20.0	36.1	4.1	3.4
Ton-km. (mill.)	11,922.0	844.6	1,205.1	276.3	420.2	865.4	51.6	2.8
	42,538.1	2,802.7	4,938.4	1,102.2	1,649.9	2,343.4	269.5	216.2

Notes: [1]Without Hessen; [2]Without Palatinate railways; [3]Including Palatinate; [4]Only standard gauge.
Top figure = 1885; Bottom figure = 1910.
Calculated from: Blum, O., et al., Verkehr u. Betrieb der Eisenbahnen. Berlin 1925.

TABLE IV

ROUTE LOST AFTER THE FIRST WORLD WAR

To France	Alsace and Lorraine	1,970 km.
	Palatinate Railway	11
To Belgium	Eupen-Malmédy*	154
To Denmark	Schleswig	254
To Poland	Provinz Posen, Danzig, Silesia, West Prussia	4,879
To Lithuania	Memelland	137
To Czechoslovakia	Hultschin (Hlučín)	31
		7,436 km.
Saar railways (until 1935)		432 km.
Running of part of Luxemburg railways		196 km.

*Including the *Eifelbahn*.

Source: Hundert Jahre deutsche Eisenbahnen, Berlin 1938, p. 44.

ROUTE LOST AFTER THE SECOND WORLD WAR

To Poland and the Soviet Union

        Oder-Neisse territories, East Prussia   9,950 km.†

†*Including narrow gauge.*

Source: Germany Reports, Federal Press and Information Office, Bonn 1955.

       Stumpf, B., Kleine Geschichte der Deutschen Eisenbahnen, Mainz 1955, p. 81.

TABLE V

RAILWAYS IN THE TWO GERMAN STATES

Deutsche Bundesbahn - German Federal Republic

	1950	1960	1970	1936
Population (mill.)	49.9	55.4	60.6	41.8
Railway length (km.)	30,840.0	30,692.0	29,479.0	30,565.0
Passengers (mill.)	1,326.2	1,543.0	1,523.0	748.7
Passenger/km. (mill.)	30,912.2	41,722.0	44,700.0	23,584.5
Tonnage (mill.)	213.4	327.2	364.5	275.6
Ton/km. (mill.)	48,078.1	63,949.0	85,052.0	46,072.1

Deutsche Reichsbahn - German Democratic Republic

	1950	1960	1970	1936
Population (mill.)	18.3	17.2	17.0	16.8
Railway length (km.)	16,174.0	16,174.0	14,909.0	14.200.0
Passengers (mill.)	954.0	943.0	636.0	761.0
Passenger/km. (mill.)	18,576.0	21,288.0	17,610.0	n.a.
Tonnage (mill.)	128.5	237.8	251.9	155.0
Ton/km. (mill.)	15,064.0	32,860.0	39,455.0	23,100.0

Source: Statistical Yearbooks of the two republics.

TABLE VI

DEUTSCHE BUNDESBAHN

	1958	1962	1966	1970	1974	1976
Route length						
Standard gauge km.	30,796	30,440	29,995	29,427	28,806	28,551
Narrow gauge km.	188	156	133	52	25	25
Electrified route km.	3,209	4,525	6,994	8,590	10,011	10,349
Passenger trains						
Expresses daily*	431	422	448	605	615	538
Semi-fast daily*	1,364	1,435	1,547	2,117	2,651	2,625
Stopping trains daily*	20,329	18,210	15,981	17,366	17,081	14,347
Average number of passengers per train	98	101	116	123	113	119
Passenger train km. (mill.)	403	376	360	395	419	388
Freight trains						
Freight trains km. (mill.)	180	191	179	216	206	183
Gross weight (t)	959	1,013	1,067	1,077	1,114	1,083
Freight trains*	12,879	13,415	13,215	13,453	12,230	10,843

* = *no. of trains run daily.*

Source: Zahlen von der Deutschen Bundesbahn, annually.

TABLE VIIa

TRAIN SERVICES IN WEST GERMANY

Daily number of fast trains (Schnellzüge)

Route	1914	1929	1962	1975
Hannover-Hamm	14	16	22	29
München-Augsburg	13	18	50	73
Hannover-Göttingen	8	10	23	46
Karlsruhe-Basel	5	12	29	32
Köln-Mainz/Wiesbaden	15	25	61	87

Source: Timetables for the respective years, Stöckl, F., Eisenbahnen in Deutschland - Von "Adler" zum "TEE", Heidelberg 1969.

TABLE VIIb

EXPRESS TRAIN SPEED IN GERMANY 1914-1938

Route	Distance (km.)	1914		1938	
		Time (min.)	Speed k.p.h.	Time (min.)	Speed k.p.h.
Halle-Nürnberg	314	270	70	250	75
Berlin-Hamburg	287	194	89	137	126
München-Würzburg	277	205	81	183	91
Berlin-Hannover	254	182	84	115	133
München-Nürnberg	199	135	88	108	111
Hannover-Hamburg	178	119	83	90	119
Leipzig-Berlin	164	116	85	76	130
Halle-Berlin	162	110	88	96	101
Salzburg-München	153	133	69	114	81
Augsburg-München	137	106	78	80	103
Osnabrück-Hannover	133	101	79	81	98
Bremen-Osnabrück	122	95	77	63	117
Leipzig-Dresden	116	87	79	82	85
Stuttgart-München	341	210	66	153	96

Source: Schlag Nach! Bibliographisches Institut, Leipzig 1938, p. 342.

TABLE VIII

THE INFLUENCE OF THE FIRST WORLD WAR ON FREIGHT TRAFFIC

(Railway freight turnover - all traffics)

Economic Area	Tonnage 1913 million tons	1913-1922 Increase (+) or decrease (-)
E. Prussia/W. Prussia	18,744	-15,584 (-83.1%) ++
Posen (Poznań)	14,008	+++
Baltic Coast	16,015	+343 (+2.1%)
Brandenburg-Berlin	40,079	+2,080 (+5.2%)
Silesia	78,404	-24,541 (-31.3%) ++++
Hamburg-Schleswig-Holstein	20,665	+2,177 (+10.5%)
Lower Saxony	41,245	+659 (+1.6%)
Central Germany	89,099	+7,228 (+8.1%)
Bavaria	27,945	+2,102 (+7.5%)
Rhineland-Westphalia	230,873	-74,634 (-32.3%)
Rhine-Main	29,683	-4,813 (-16.2%)
Baden-Württemberg	39,939	+1,206 (+3.0%)
Saar	22,490	+++
Alsace-Lorraine	38,022	+++

++ = estimate to allow for territorial change.
+++ = territory completely lost to Germany: in case of the Saar until 1935.
++++ = Upper and Lower Silesia. Estimate to allow for territorial change.

Source: Calculated from Tiessen, E., Der Gesamt-Güterverkehr auf den deutschen Eisenbahnen 1913-1922, Berlin 1925; and Blum, O., et al., Verkehr und Betrieb der Eisenbahnen, Berlin 1925.

NOTES

1. The historical background of the period is portrayed in the classical work of H. von Treitschke, *Deutsche Geschichte im 19 Jahrhundert*, I-VI, Leipzig 1889-1894.

2. The leading Bavarian civil servant was J.R. von Baader, a keen advocate of railways and a force behind the Nürnberg-Fürth railway. His main opposition came from the Bavarian King, an equally keen supporter of canals and the main force behind the abortive *Ludwigskanal* (1836-1845), an unsuccessful link between the Main and Danube.

3. At this time (1830's) many proposals were being made and little agreement in detail exists among the histories of the period. Among the railways proposed were lines from Wesel to Rheine (Rhine-Ems link) and Elberfeld to Rheine (one of Harkort's proposals as a Ruhr-Ems link). The purpose of all was, however, clearly to avoid using the Dutch-controlled Rhine mouth for German sea-going trade. Even earlier, in 1824, a senior civil servant in Braunschweig, von Amsberg, had proposed a railway from the Hanseatic towns of Hamburg and Bremen via Hannover and Braunschweig to the south of Germany.

4. List, F., *Über ein sächsisches Eisenbahnsystem als Grundlage eines allgemeinen Eisenbahnsystems*, Leipzig 1833. See Hirst, M.E., *The Life of Freidrich List*, London 1909.

5. The Erkrath-Hochdahl incline (1:30) was worked by rope until 1927. The working was then done by banking engines, but these were abandoned after electrification in the early 1960's. By making the line just over one kilometre longer, the bank could have been avoided. Plans to do this before the 1914-18 War never materialised. See Schneider, A., *Gebirgsbahnen Europas*, Zürich 1963, pp. 347-349.

6. Originally there was no physical connection at Friedrichsfeld, since the Main-Neckar Railway was 1,435 mm. gauge and the Mannheim-Heidelberg line 1,600 mm. gauge (Baden State Railway).

7. In 1879 at the Berlin Trade Exhibition, Werner von Siemens had displayed a small electric locomotive that drew a short train of passengers round part of the stands.

8. Julius Dorpmüller (1869-1945). Entered service with the Prussian State Railways, 1893; 1922 President of the Railway Directorate Oppeln; 1924 President of the Railway Directorate Essen; 1925 Deputy Director of the *Deutsche Reichsbahn*; 1926 General Director of the *Deutsche Reichsbahn*; 1937 Reichs Transport Minister. In May 1945, he was asked by the Western allies to reorganise the German railways, but died in July of the same year.

9. It is seldom realised that by 1926 Germany had the densest civil airline system in Europe, with regular flights serving all the principal towns.

10. The Deutsche Bundesbahn offers special tariffs in the zonal border planning region (Zonenrandgebiet) for all places whose railway charges have been substantially increased because of the new Inter-German border cutting across pre-1945 routes.

11. Track dismantled in Germany was reputedly used to build the Trans-Mongol Railway from Nerchinsk (U.S.S.R.) to Ulan-Bator.

12. The Germans developed the change of gauge vehicle before the Second World War. Since 1945, most change of gauge coaches and wagons for use between the Soviet Union and the rest of Europe have been manufactured or designed in the Halle-Ammendorf works in the G.D.R.

13. Plans for the future of the Deutsche Bundesbahn have been discussed in: Oeftering, H.M., Die Konzentrationsmassnahmen der DB und die Sicherstellung einer optimalen Verkehrsbedienung in der Fläche, Bundesbahn 7, 1966, pp. 1-9; Bundesminister für Verkehr, Zielsetzungen des BMV für die Unternehumgspolitik der Deutschen Bundesbahn und für den Öffentlichen Personenverkehr, Schriftenreihe des BMV 29, 1975, also Verkehrspolitik '76 - Grundsatzprobleme und Schwerpunkte, Schriftenreihe des BMV 50, 1976; Gaebel, D., Volkswirtschaftliche Bedeutung in DB in unserer Gesellschaft, Frankfurt am Main 1975.

14. Union Internationale des Chemins de Fer, Plan Directeur du Chemin de Fer Européen de l'Avenir, Paris 1973.

15. Reputedly in the late 1940's it commonly took longer to go by rail from Berlin to Leipzig or Dresden than it took by coach in the eighteenth century.

BIBLIOGRAPHY

AUTORENKOLLEKTIV. *Uns gehören die Schienenwege - Festschrift zum 125 jährigen Jubiläum der Eisenbahnen in Deutschland.* Berlin East 1961.

BERKENDORF, P., FORSTHOFF, E. 'Die gemeinwirtschaftliche Verkehrsbedienung der Deutschen Bundesbahn'. *Schriftenreihe "Die Bundesbahn", Folge 9,* Darmstadt 1958.

BLUM, O. *Verkehrsgeographie.* Berlin 1936.

BLUM, O., et al. 'Verkehr und Betrieb der Eisenbahnen. *Handbibliothek für Bauingenieure, 8,* Braunschweig 1925.

BLUM, O. 'Grundsätzliches zum Ausbau des Verkehrsnetzes'. *Wirtschaftsdienst, 12,* 1927, pp. 541-543.

BORN, E. *Hundert Jahre bayrische Ostbahn.* München 1957.

BORN, E. 'Lokomotiven und Wagen der deutschen Eisenbahnen'. *Kleine Eisenbahn-Bücherei, Band 2,* Mainz 1958.

BORN, E. *Pioniere des Eisenbahnwesens.* Darmstadt 1964.

BÖTTCHER, W. 'Die Berlin-Stettiner Eisenbahn'. *Kleine Eisenbahnschriften, 19,* Dortmund n.d.

BÖTTCHER, W. 'Die Nordbrabant-Deutsche Eisenbahn und ihre Lokomotiven'. *Kleine Eisenbahnschriften, 12,* Dortmund n.d.

BÖTTCHER, W. 'Die Berlin-Hamburger Eisenbahn'. *Kleine Eisenbahnschriften, 24,* Dortmund n.d.

BÖTTCHER, W. 'Die Berlin-Potsdam-Magdeburger Eisenbahn'. *Kleine Eisenbahnschriften, 11,* Dortmund n.d.

BÖTTCHER, W. 'Auf deutschen Eisenbahnen anno dazumal
1. Kampf der preussischen Regierung um die Einführung von Nachtzügen
2. Die Niederschlesisch-Märkische Eisenbahn'.
Kleine Eisenbahnschriften, 1, Dortmund n.d.

BUCHHOLZ, H.J. *Der Eisenbahnverkehrsknoten Hamm (Westf.) - Entwicklung und Wandlung seiner Bedeutung.* Hamm 1976.

BUNDESBAHNDIREKTION ESSEN. *S-Bahn Ruhrgebiet - Pressekonferenz Mai 1974.* Essen 1974.

DAVIES, W.J.K. 'West Germany'. *Continental Railway Handbooks.* London 1971.

DELVENDAHL, H.   'Die Ergänzungsstrecken Köln-Gross Gerau und Hannover-
    Gemünden nach dem Ausbauprogramm'.   *Bundesbahn, 7-8,* 1971,
    pp. 315-318.

DEUTSCHE BUNDESBAHN.   *Eisenbahn-Verkehrsdienst.*   Darmstadt 1961.

DEUTSCHE BUNDESBAHN.   *Die Deutsche Bundesbahn in unserer Gesellschaft.*
    Frankfurt/Main 1975.

DEUTSCHE BUNDESBAHN.   *Vom Adler zum Komet - Bilder zur Geschichte
    der deutschen Eisenbahn.*   Bonn 1956.

DITT, H., SCHÖLLER, P.   'Die Entwicklung des Eisenbahnnetzes in
    Nordwestdeutschland'.   *Westfälische Forschungen, 8,* 1955,
    pp. 150-180.

DORPMÜLLER, J.   *Hundert Jahre Deutsche Eisenbahnen.*   Leipzig 1935
    (2nd ed. 1938).

DÖRSCHEL, W. (ed.)   Verkehrsgeographie - *Lehrbuch für die Berufsbildung
    der Deutschen Reichsbahn, IV,* Berlin E. 1968.

DOST, P.   *Zur Geschichte der Warschau-Wiener Eisenbahn.*   Dortmund
    1965.

DOST, P.   *Der rote Teppich - Geschichte der Staatszüge und Salonwagen.*
    Stuttgart 1965.

DOST, P.   'Die K.M.E. und andere Militäreisenbahnen I-V'.
    Kleine Eisenbahnschriften, 13-17, Dortmund n.d.

DUBROWSKY, H.J.   *Die Zusammenarbeit der RGW-Länder auf dem Gebiet
    des Transportwesens.*   Berlin E. 1975.

ENGELBRACHT, T.H.   'Kartographische Darstellung der Güterbewegung
    auf deutschen Eisenbahnen'.   *Petermanns Mitteilungen, 69,*
    1923, pp. 249-250.

ERNST, F.   *Rheingold - Luxuszug durch fünf Jahrzehnte.*   Düsseldorf
    1971.

FAKINER, F.   'Der moderne Personenbahnhof in Technik und
    Betriebsweise'.   *Kleine Eisenbahn-Bücherei, Band 3,*
    Mainz 1959.

FERRARIUS.   'Das deutsche Bahnnetz 1870 und 1911'.   *Petermanns
    Mitteilungen, 57,* 1911, pp. 323-330.

FUCHS, K.   *Die Erschliessung des Siegerlandes durch die Eisenbahn.*
    Wiesbaden 1969.

FÜRST, A.   *Die Welt auf Schienen.*   München 1925.

GIESE, K.   'Die deutschen Städte und das Eisenbahnwesen', in *Das
    Deutsche Eisenbahnwesen der Gegenwart.*   Berlin 1927.

GLÄSSEL, D.   'Zur Eisenbahn-Geschichte Schlesiens'.   *Kleine Eisenbahnschriften, 43,* Dortmund n.d.

GOTTWALDT, A.B.   *Deutsche Reichsbahn 1935 - Ein Text und Bildreport.*   Stuttgart 1975.

GROLL, A., KAYSER, H.   'Fahrplanwesen'.   *Eisenbahn-Lehrbücherei der Deutschen Bundesbahn, Band 211,* Starnberg 1966.

HASSERT, K.   *Allgemeine Verkehrsgeographie I + II.*   Berlin 1931.

HAUFE, H.   'Die geographische Struktur des deutschen Eisenbahnverkehrs'. *Veröffentlichung des Geog. Seminars d. Universität Leipzig, 2,* 1931.

HEMBERGER, A.   *Die Eisenbahnfibel.*   Bamberg 1954.

HOFFMANN, R.   'Rückzug der Eisenbahn aus der Fläche? Ein Problem der Regional- und Verkehrspolitik'.   *Akademie für Raumforschung und Landesplanung, Band 46,* Hannover 1965.

HOTTES, K.H.   'Verkehrsgeographischer Strukturwandel im Rhein-Ruhrgebiet'.   *Geographisches Taschenbuch,* 1970-1972, pp. 102-114.

HÜTTMANN, E.   *Verkehrsgeographische Probleme am Beispiel der Eisenbahnen Schleswig-Holsteins.*   Hamburg 1949.

JOHN, G.   'Die Verkehrsströme innerhalb der BRD nach Gütergruppen und Verkehrsarten'.   *Deut. Inst. für Wirtschaftsforschung - Beiträge zur Strukturforschung, 3,* Berlin 1967.

KELLER, E.   'Die verkehrsgeographischen Grundlagen der deutschen Eisenbahnumwege'.   *Archiv für Eisenabhnwesen, 52,* 1929, pp. 343-362, 583-630.

KOBSCHÄTZKY, H.   *Streckenatlas der deutschen Eisenbahnen 1835-1892.*   Düsseldorf 1972.

KRAUS, T.   'Verkehrsgeographische Betrachtungen über die Eisenbahnen in den Grenzgebieten Mittel- und Osteuropas'.   *Geographische Zeitschrift, 31,* 1925, pp. 221-236.

KUMBIER, G.   'Die Bedienung des Güterverkehrs im Ruhrgebiet durch die Deutsche Bundesbahn'.   *Akademie für Raumforschung und Landesplanung, 12,* 1959, pp. 125-131.

KUNTZEMÜLLER, A.   *Die Badischen Eisenbahnen.*   Karlsruhe 1953.

KURZE, J.   *Lebensadern der Wirtschaft.*   Bonn 1959.

KURZE, J., et al.   '1835-1960 - 125 Jahre Deutsche Eisenbahnen'. *Die Bundesbahn - 21/22 Sonderausgabe,* 1960.

KURZE, J. (ed.)   *Zehn Jahre Wiederaufbau bei der Deutschen Bundesbahn.*
    Darmstadt 1955.

LAEMMERHOLD, F.   'DB-Ausbauprogramm und Bundesverkehrswegplanung'.
    Die Bundesbahn, 7/8, 1971, pp. 315-318.

   'L'Allemagne - République Fédérale'.   *La Vie du Rail, 1472,*
    1974.

LASSMANN, J.E.   *Der Eisenbahnkrieg - Taktische Studie.*   Berlin 1867.

LENZ, F.   *Friedrich List.*   Berlin 1936.

LOGEMANN, W.   'Vor hundert Jahren (1860): ein geschichtlicher
    Rückblick auf die 1860 neu eröffneten Eisenbahnstrecken'.
    Die Bundesbahn, 34, 1960, pp. 1220-1236.

LOGEMANN, W.   'Hundert Jahre Nord-Süd-Verbindung durch das Leinetal
    ("Hannoversche Südbahn")'.   *Die Bundesbahn, 30,* 1956,
    pp. 1425-1433.

LOGEMANN, W.   'Die "unechte" Weserbahn als Entlastungslinie für die
    stark beanspruchten Nord-Süd-Strecken der Deutschen
    Bundesbahn'.   *Eisenbahntechnische Rundschau, 3,* 1954,
    pp. 527-551.

LOTZ, W.   'Verkehrsentwicklung in Deutschland 1800-1900'.   *Aus
    Natur- und Geisteswelt, 15,* Berlin 1906.

MAEDEL, K.E.   *Liebe alte Bimmelbahn - eine Erinnerung an unsere
    deutschen Klein- und Nebenbahnen.*   Stuttgart 1967.

MAEDEL, K.E.   *Weite Welt des Schienenstrangs.*   Stuttgart 1965.

   Das Eisenbahn-Jahrhundert.   Stuttgart 1973.

   Die deutschen Dampflokomotiven gestern und heute.
    Berlin E. 1962.

MARANDON, J.C.   'Der kombinierte Güterverkehr Schiene/Strasse in
    der BRD als Faktor der Industrieansiedlung'.
    Materialien zur Raumforschung, VI, Bochum 1973.

MELLOR, R.E.H.   'German Railway Reconstruction 1945-1952'.
    Railway World, 14, 1953, pp. 187-190, 208-211.

MOLLOWO, H.J.   'Die Lokalbahnen im Steigerwald und in der Fränkischen
    Alb'.   *Mitteilungen d. Fränkischen Geog. Gesellschaft, 19,*
    Erlangen 1972, pp. 237-257.

MÜHL, A., SEIDEL, K.   *Die Württembergischen Staatseisenbahnen.*
    Stuttgart 1975.

NAVAL INTELLIGENCE DIVISION. 'Germany: IV - Ports and Communications'. *Geographical Handbook Series*, London 1945.

OBERMEYER, H.J. 'Bahnanlagen und Fahrdienst'. *Taschenbuch der Eisenbahn, 2*, Stuttgart 1977.

O'DELL, A.C. *Railways and Geography*. London 1956. (Revised second edition, with Richards, P.S., 1966).

OLBRICH, P. 'Betrieb und Verkehr bei der "Deutschen Reichsbahn" in der sowjetischen Besatzungszone'. *Materialien zur Wirtschaftslage in der sowjetischen Zone*, Bonn 1957.

OTTMANN, K. 'Berlin und seine Eisenbahnen bis zur Jahrhundertwende'. *Archiv des Eisenbahnwesens, 74*, 1964, pp. 293-317.

POTTGIESSER, H. 'Die Reichsbahn im Ostfeldzug'. *Die Wehrmacht im Kampf, 26*, Neckargemünd 1960.

REICHSVERKEHRSMINISTERIUM. *Die Deutschen Eisenbahnen 1910-1920*. Berlin 1923.

ROSSBERG, R.R. *Tempo 200 - Eisenbahn heute*. Stuttgart 1971.

RUDNICK, K. 'Organisation der Deutschen Bundesbahn'. *Eisenbahn-Lehrbücherei der Deutschen Bundesbahn, Band 3*, Starnberg 1964.

RUFF, B. *Die Höllentalbahn - aus Vergangenheit in die Gegenwart*. Stuttgart 1970.

RUTZ, W. 'Die Alpenquerungen - ihre Verkehrsbeziehungen, Verkehrsbedeutung und Ausnutzung durch Verkehrswege'. *Nürnberger Wirtschafts- und Sozialgeographische Arbeiten, 10*, 1969.

SCHMITZ, A. 'Verkehrsgeographie'. *Eisenbahn-Lehrbücherei der Deutschen Bundesbahn, Band 61*, Starnberg 1968.

SCHNEIDER, A. *Gebirgsbahnen Europas*. Zürich 1963.

SCHUCHARDT, A.G. *50 Jahre Leipzig-Hauptbahnhof*. Berlin E. 1965.

SCHULZ, F.T. *Die Ludwigsbahn - die erste deutsche Eisenbahn*. Leipzig 1935.

SCHULTZ-RHONDORF, F.C. 'Die Verkehrsströme der Kohle im Raum der BRD zwischen 1913 und 1957'. *Forschungen zur deutschen Landeskunde, 146*, 1964.

Sechs Jahrzehnte Köln-Bonner Eisenbahn. Koln 1956.

SEEBOHM, H.C. *Die Verkehrswege in der Bundesrepublik Deutschland*. München 1964.

SEEBOHM, H.C.   'Verkehrspolitik 1949-1965'. *Schriftenreihe des Bundesministers für Verkehr, 29,* Bonn 1965.

SEIDEL, W.   'Die Eisenbahn in der Sowjetzone'. *Materialien zur Wirtschaftslage in der sowjetischen Zone,* Bonn 1954.

SÖLCH, W.   <u>Orient-Express</u>: *Glanzzeit und Niedergang eines Luxuszuges.* Düsseldorf 1974.

STÖCKL, F.   *Eisenbahnen in Deutschland - Vom <u>Adler</u> zum <u>TEE</u>.* Heidelberg 1969.

STÖCKL, F.   'Rollende Hotels - die Internationale Schlafwagengesellschaft'. *Eisenbahnen der Erde, VIII,* Wien 1967.

STÖCKL, F.   'Deutschland'. *Eisenbahnen der Erde, VI,* Wien 1964.

STÖCKL, F.   *Europäische Eisenbahnzüge mit klangvollen Namen.* Darmstadt 1958.

STÖCKL, F.   'Verkehr, Politik und Vernunft'. *Eisenbahn, 6,* 1955, pp. 99-103.

STÖCKL, F., JEANMAIRE, C.   *Komfort auf Schienen - Schlafwagen, Speisewagen, Salonwagen der europäischen Eisenbahnen.* Basel 1970.

STUMPF, B.   'Kleine Geschichte der deutschen Eisenbahnen'. *Kleine Eisenbahn-Bücherei, Band 1,* Mainz 1955.

STÜRMER, D.   'Die Eisenbahnen Deutschlands'. *Petermanns Mitteilungen, 33,* 1878, pp. 170-175.

STÜRMER, D.   *Geschichte der Eisenbahnen: Entwicklung und jetzige Gestaltung sämtlicher Eisenbahnnetze der Erde.* Bromberg 1872.

SÜSS, W.   *Die Geschichte des Münchner Hauptbahnhofes.* München 1954.

TEUBERT, W.   'Die Bedeutung der verschiedenen Verkehrsmittel in Deutschland'. *Comptes rendus du Congrès Internationale de Géographie, Tome II,* 1938, pp. 137-158, Leiden 1938.

TIESSEN, E.   'Der Gesamt-Güterverkehr auf den deutschen Eisenbahnen 1913-1922'. *Grundkarten der deutschen Wirtschaft,* Berlin 1925.

ULBRICHT, J.F., RITZAU, H.J.   *Deutsche Eisenbahn-Geschichte II - Sachsen.* Landsberg 1977.

UNION INTERNATIONALE DES CHEMINS DE FER.   *Plan Directeur du Chemin de Fer Européen de l'Avenir.* Paris 1973.

VÖLKER, W. 'Die Entwicklung der Eisenbahnen im Ruhrgebiet - ihre Bedeutung und Aufgaben für Wirtschaft und Industrie'. *Akademie für Raumforschung und Landesplanung, 12,* 1959, pp. 101-123.

'Von der Köln-Mindener Eisenbahn zur Rhein-Ruhr S-Bahn'. *Moderne Eisenbahn, 27,* 1967, pp. 10-13.

VON HEDEMANN, H. 'Schnellzugskarte des Deutschen Reiches'. *Petermanns Mitteilungen, 60,* 1914, p. 30.

VON MEYER, A. *Geschichte und Geographie der deutschen Eisenbahnen von ihrer Entstehung bis auf die Gegenwart.* Berlin 1891.

WACHTEL, F., et al. 'Die Entwicklung der Eisenbahnen in der ehemaligen französischen Besatzungszone'. *Jahrbuch für Eisenbahnwesen,* Köln 1951.

WENZEL, H. *Die Südwestdeutschen Eisenbahnen in der französischen Zone (S.W.D.E.).* Mainz 1976.

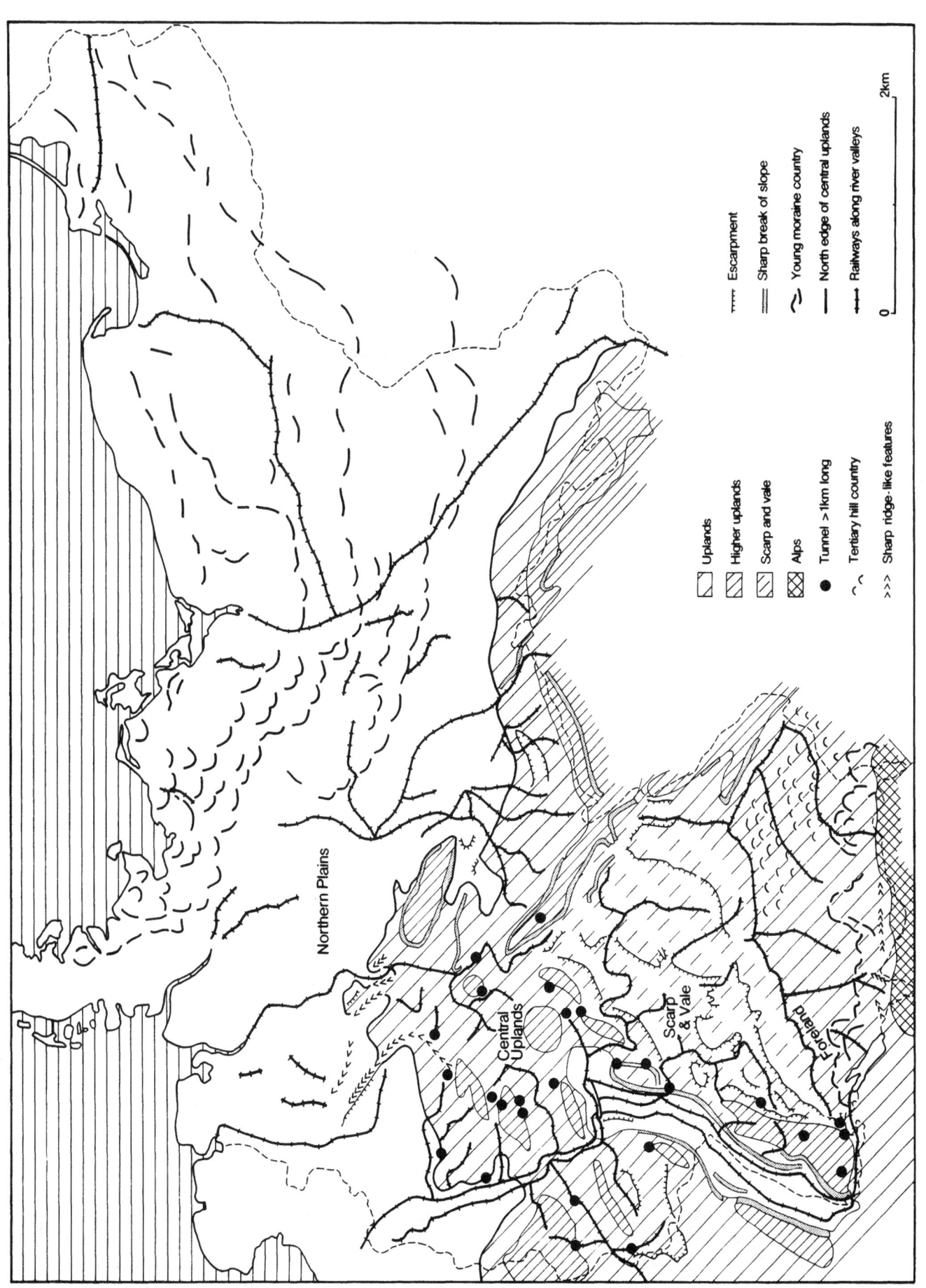

1 Physical setting for railways. River valleys form a significant influence on the alignment of many railways. Frontiers as in 1914.

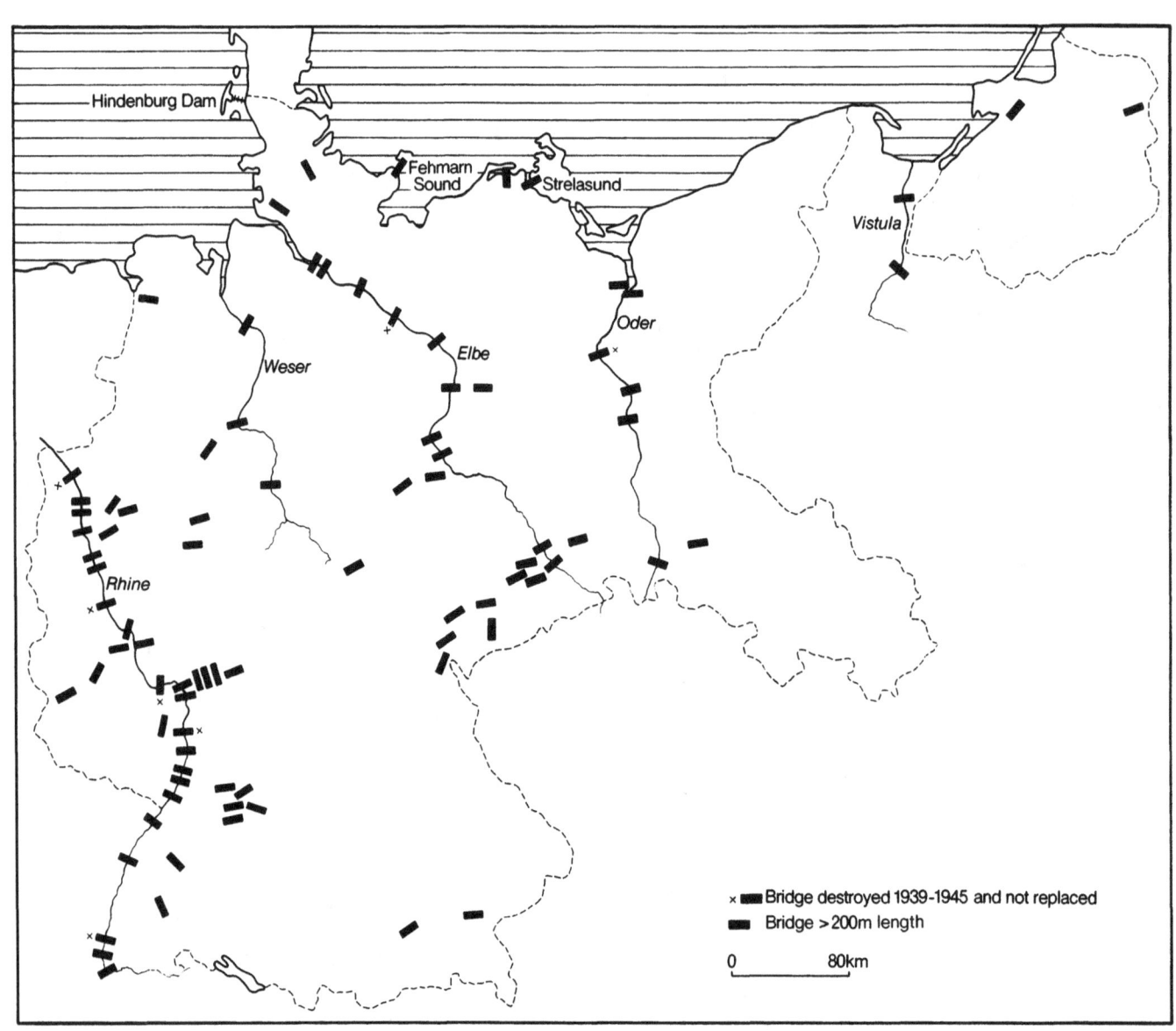

2 Major railway bridges. The main rivers - the Rhine, Weser, Elbe and Oder - constitute considerable obstacles. Several Rhine bridges of purely strategic significance were not rebuilt after 1945, while one Elbe bridge across the boundary between the British and Soviet occupation zones was not replaced. The rough terrain in the uplands of Saxony and Thüringen is marked by several important bridges. Frontiers as in 1937.
Source: various.

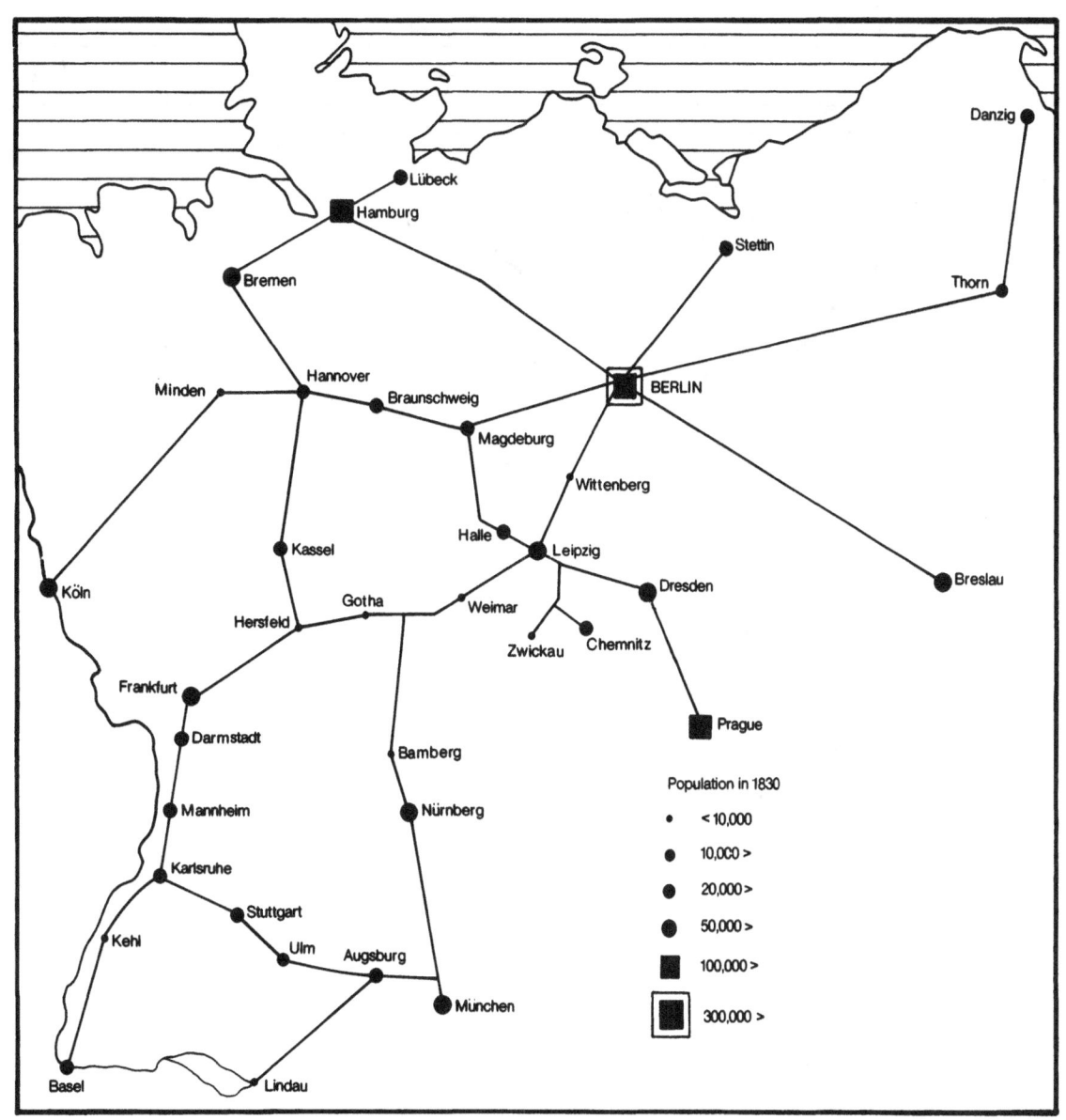

3 Friedrich List's proposals for an All-German railway system (1833).
  <u>Source</u>: facsimile reproduction of the original map.

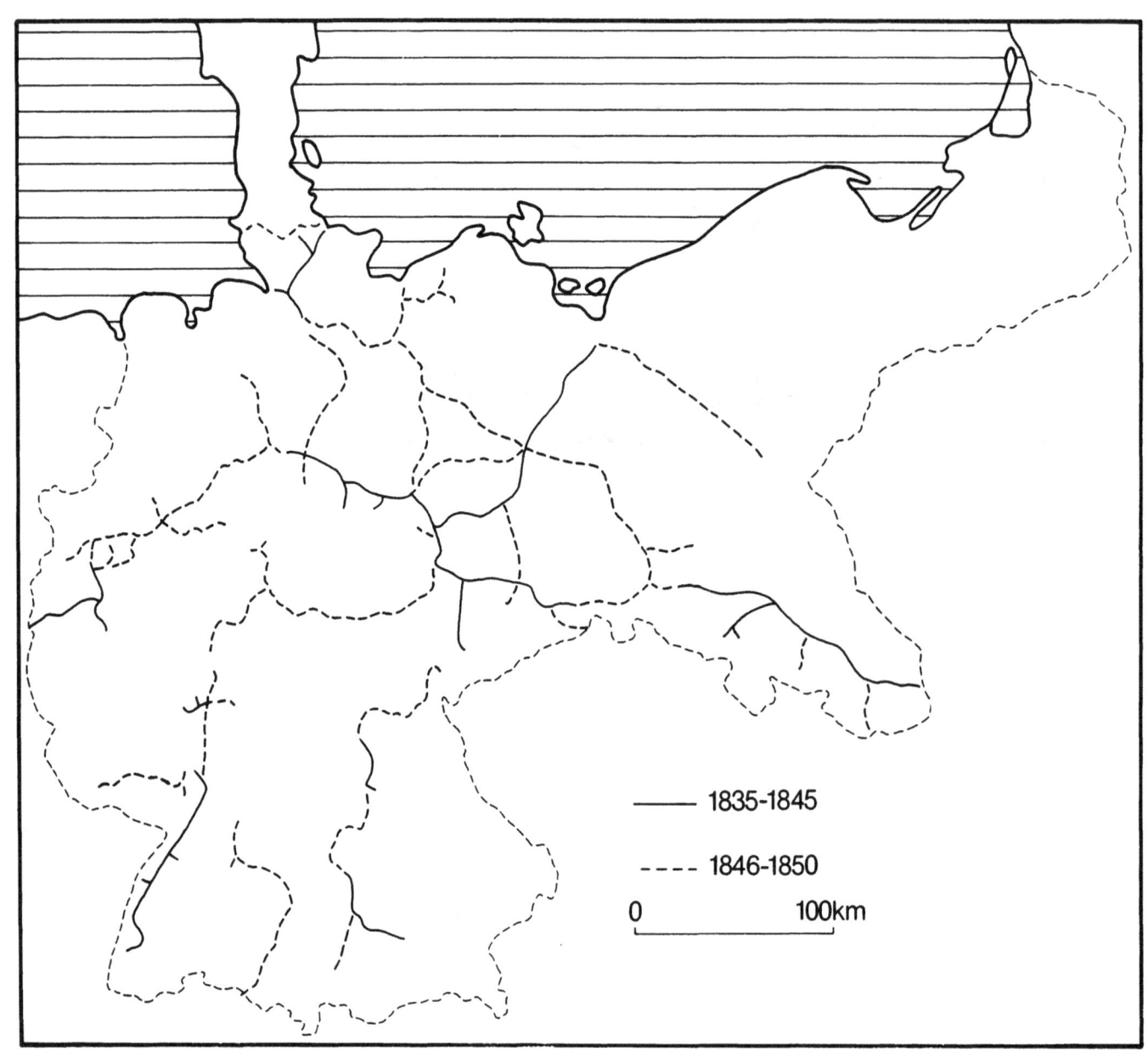

4 Growth of railway route - I. The early importance of the main east-west and north-south axes of movement is beginning to appear.
Source: <u>Hundert Jahre Deutsche Eisenbahnen</u>, Berlin 1938.

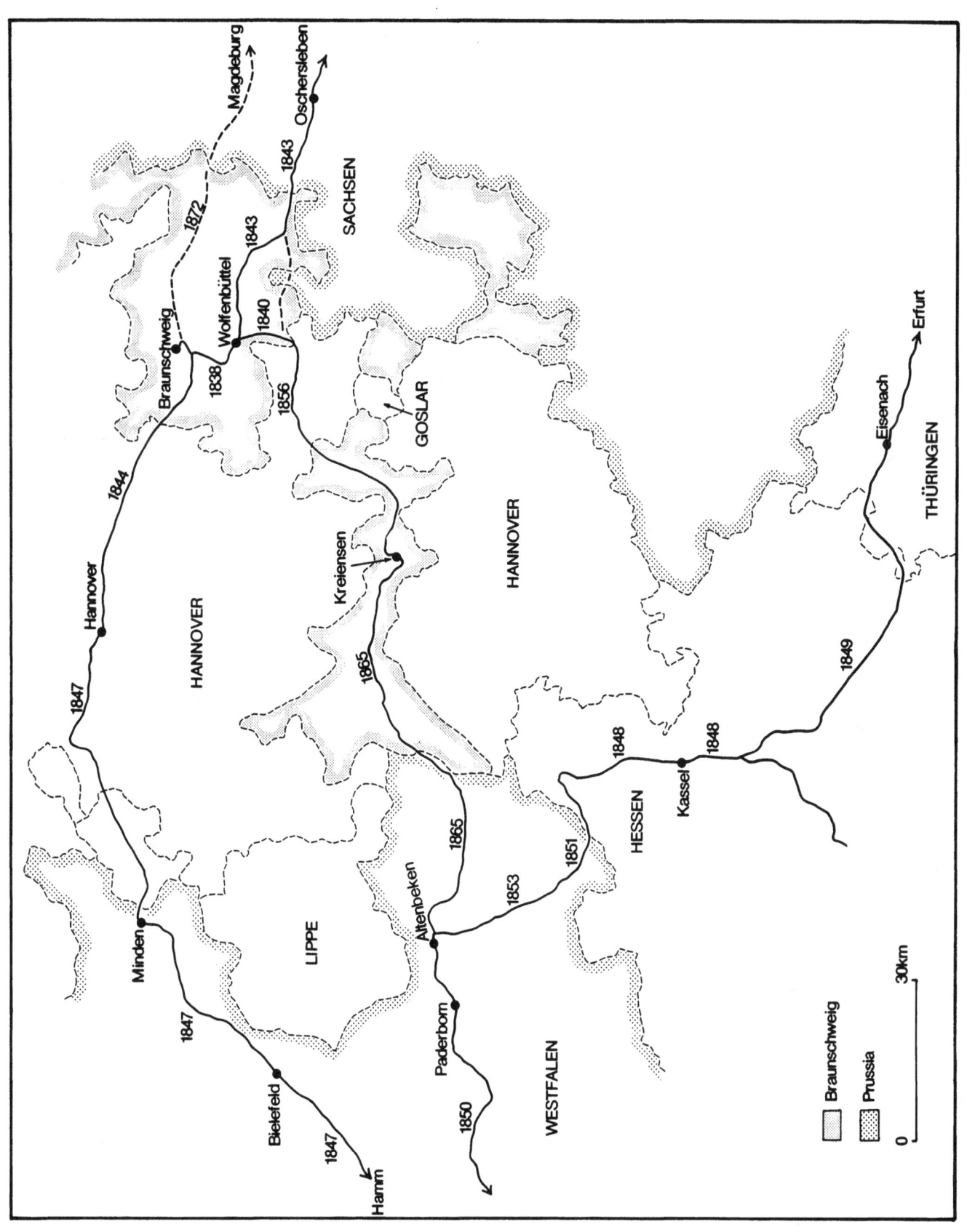

5 Prussian railway connections to the west. Prussia sought to join its recently acquired western territories to Berlin across relatively friendly princedoms as relations with the Kingdom of Hannover deteriorated.
Source: various including Kobschätzky, H., <u>Streckenatlas der Deutschen Eisenbahnen</u> 1835-1892, Düsseldorf 1971.

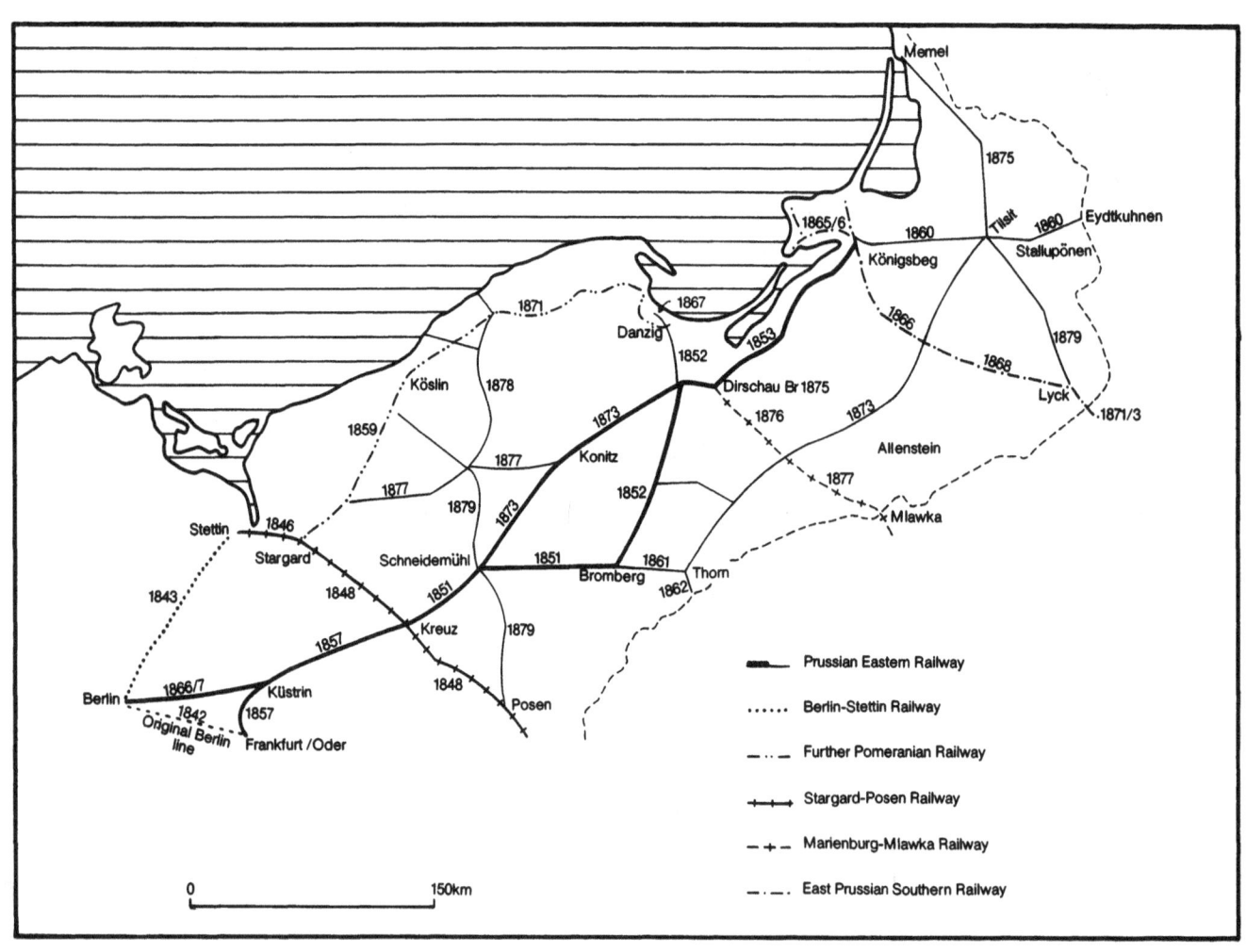

6 The Prussian Eastern Railway. This was a strategically vital route that emerged piecemeal by adding several short cuts. From Berlin, the first route was via Stettin; then via Frankfurt/Oder; this was followed by the Küstrin cut-off; and finally the short route via Konitz was completed.
Source: as for Fig. 5.

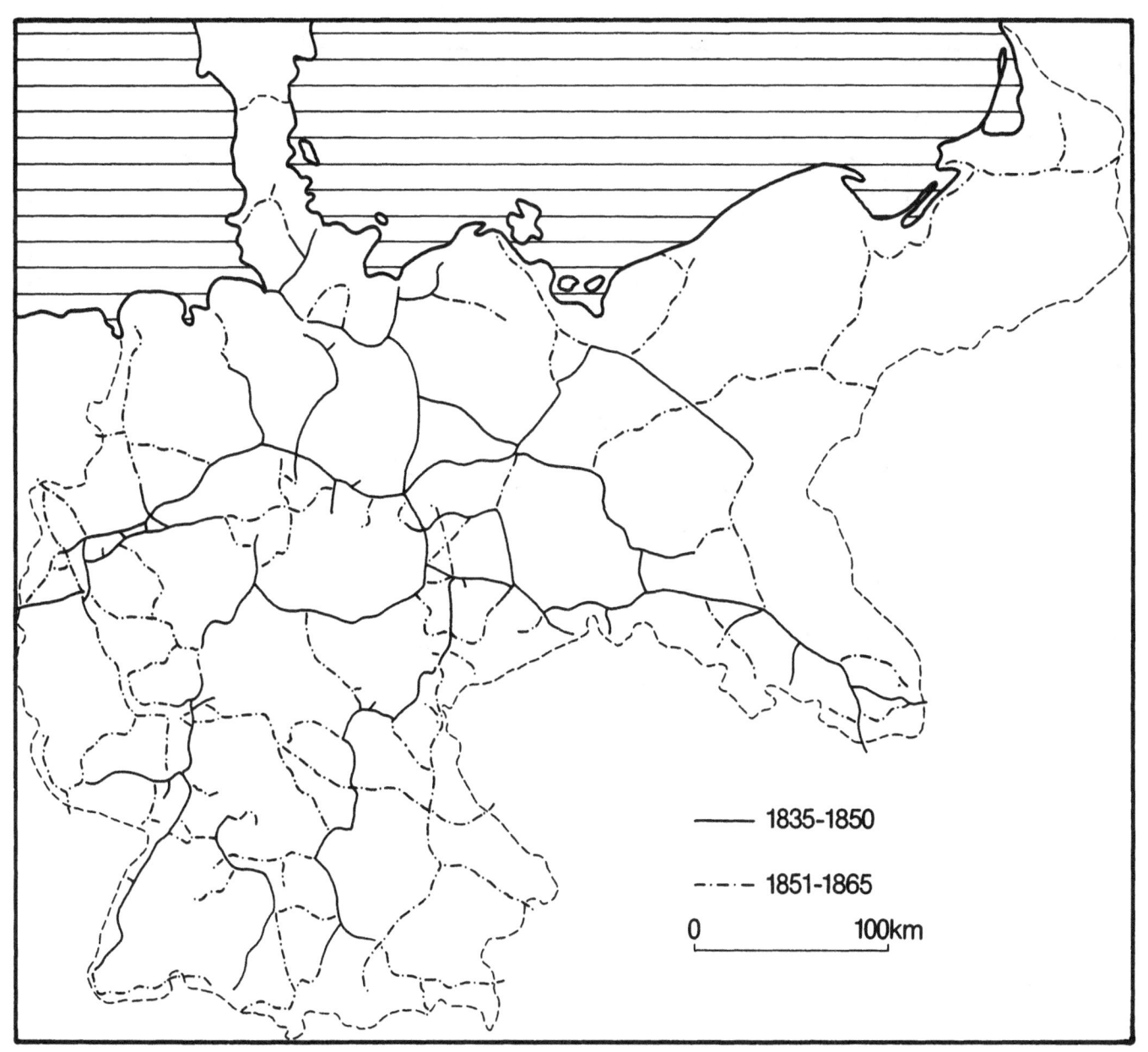

7 Growth of railway route - II. The major features of the network were in all essentials complete by 1865.
Source: as for Fig. 4.

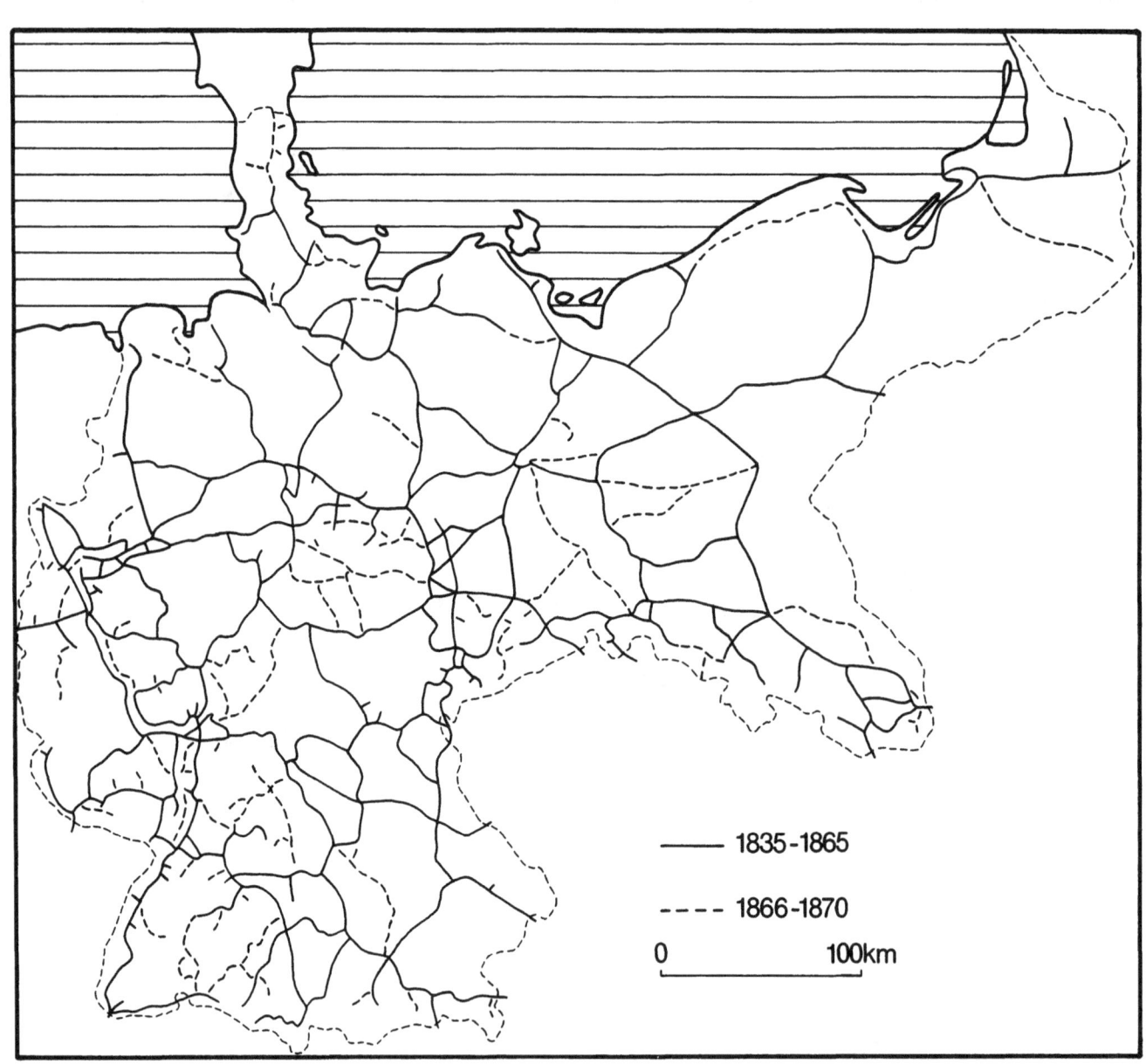

8 Growth of railway route - III. The construction of railways was now turning to secondary mainlines.
Source: as for Fig. 4.

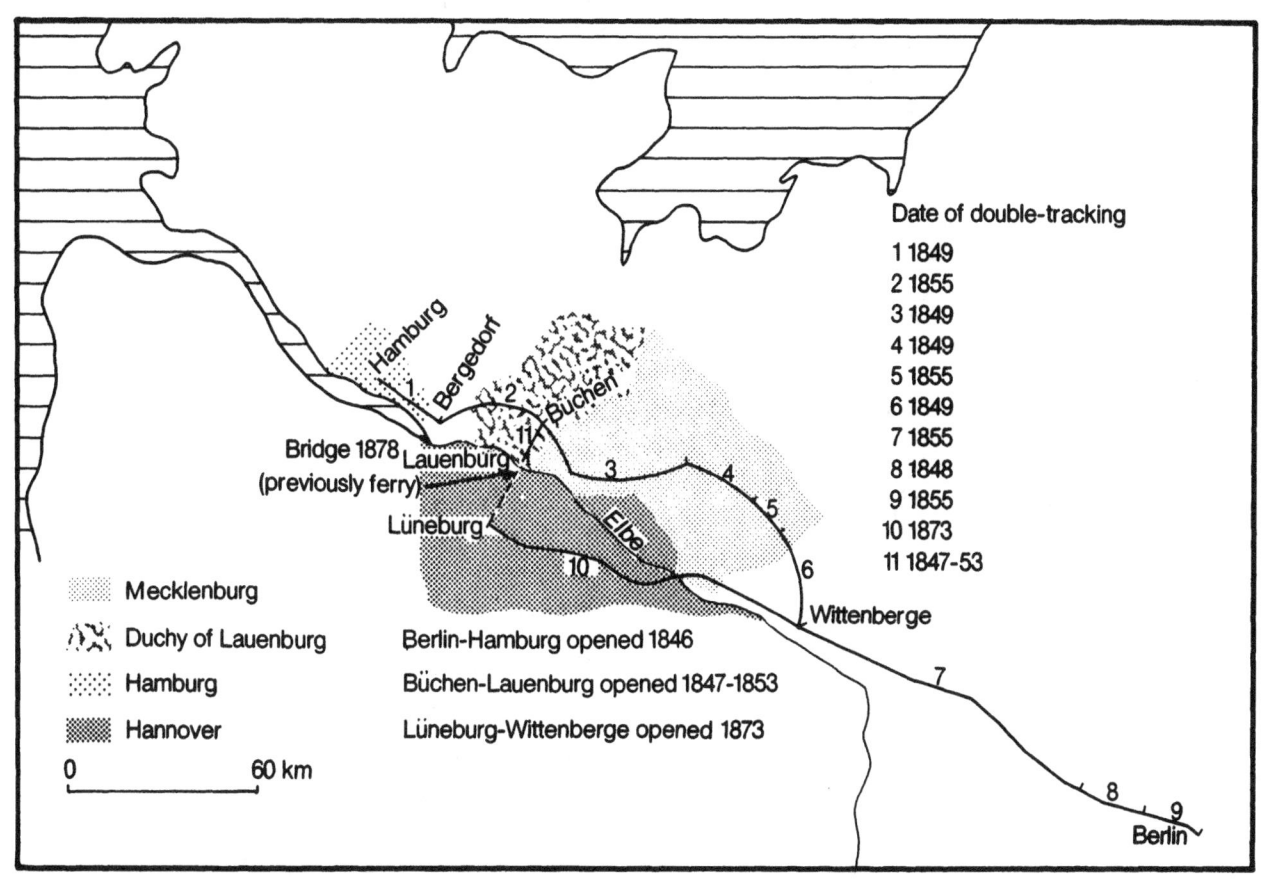

9 The Berlin-Hamburg Railway. This was an early exercise in overcoming the <u>Kleinstaaterei</u> of the period, especially the clash of interests between Mecklenburg, Lauenburg (under the Danish king), Hannover, Bergedorf and Hamburg.
<u>Source</u>: as for Fig. 5.

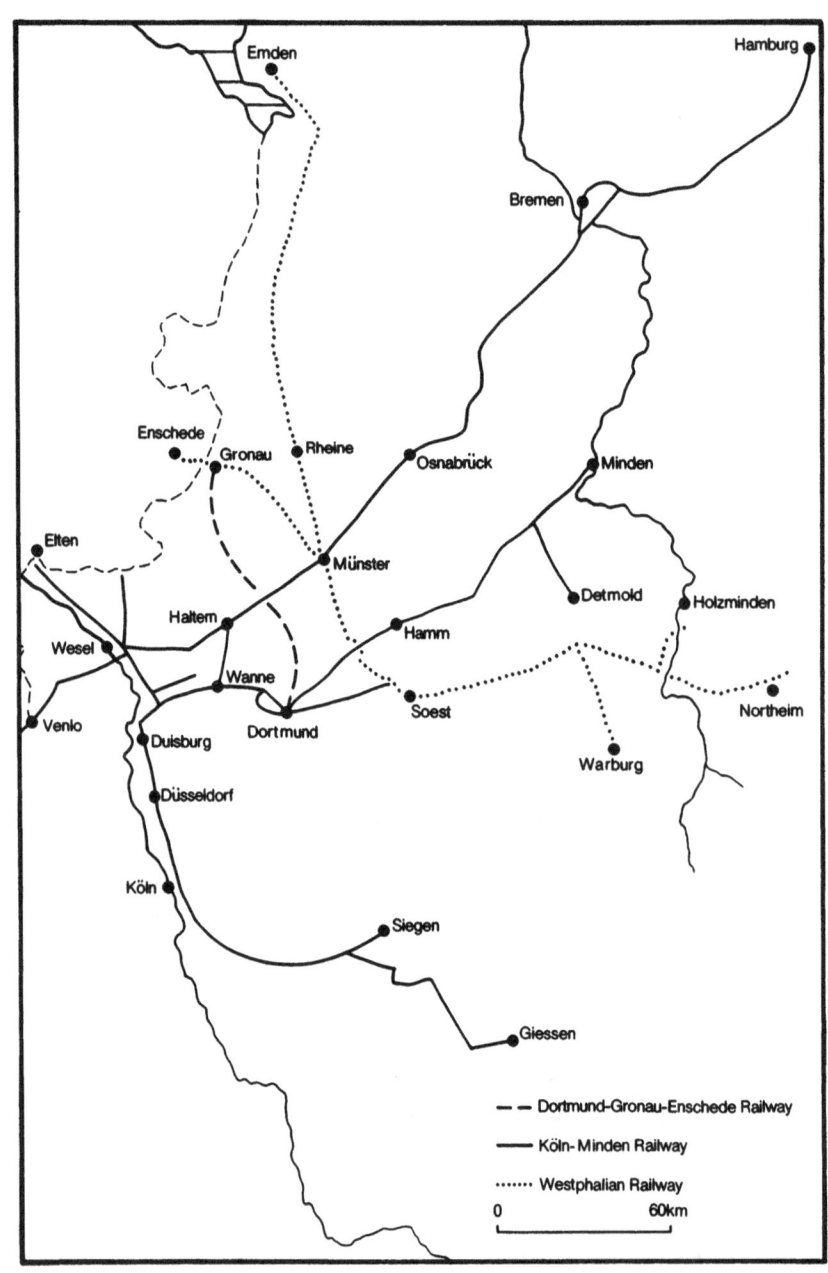

10 The Köln-Minden Railway. One of the early major projects, this railway took over completion of the abortive Hamburg-Venlo-Paris Railway and had vacillating relations with the Westphalian Railway and the short Dortmund-Gronau-Enschede Railway.
Source: as for Fig. 5.

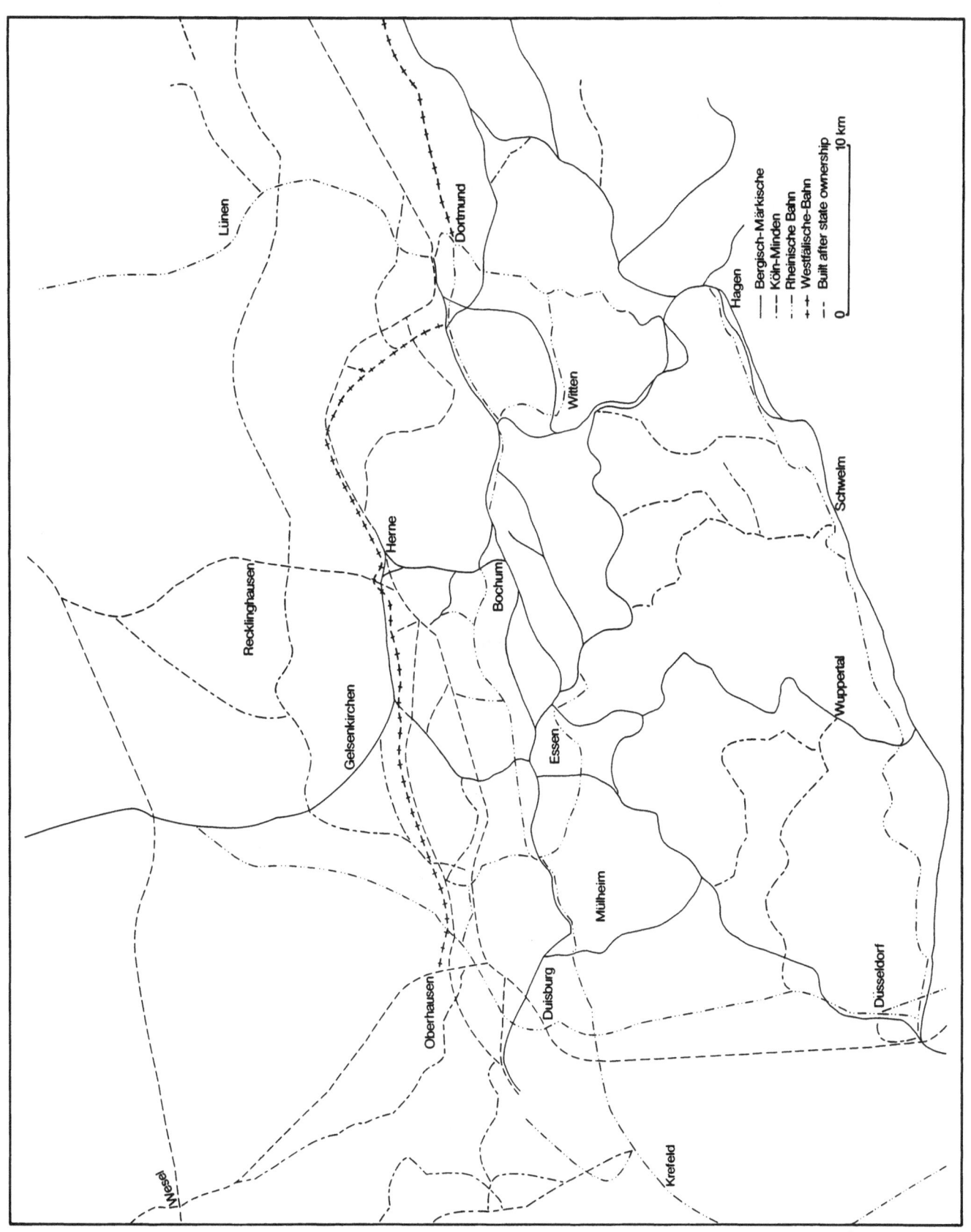

11 Railways in the Ruhr before state ownership. This was one of the
most complex areas of different railway companies in Germany before
the extension of full state ownership under the Prussian state
railways. The large and valuable coal and iron traffic was the
main attraction.
Source: as for Fig. 5.

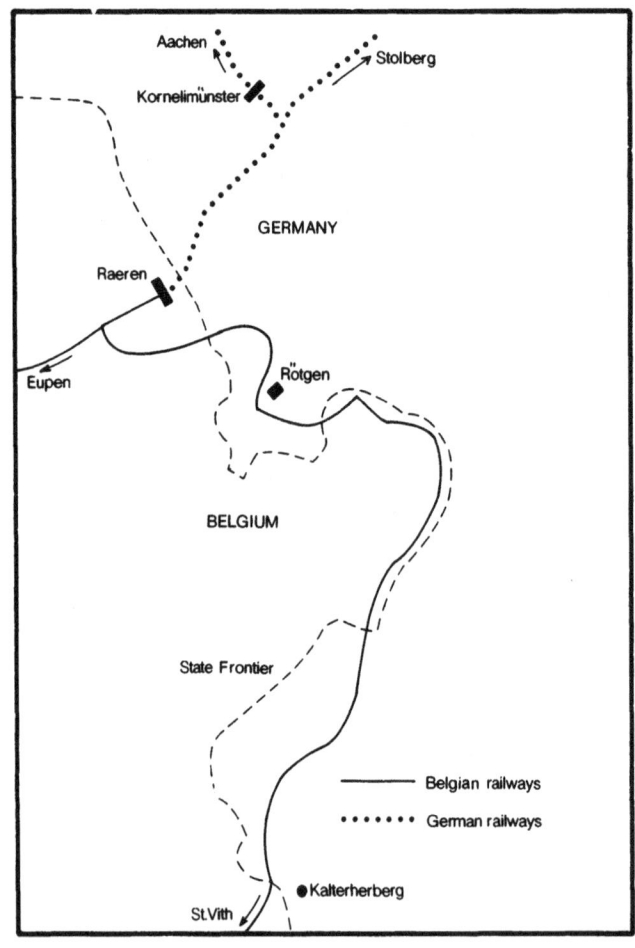

12 The <u>Eifelbahn</u>. When Belgium took over German Eupen and Malmédy
after the first world war, this German railway was incorporated
into the Belgian state railways, although much of the line ran through
remaining German territory. Special and complex operating arrangements
were made for through trains from Belgium back into Belgium.

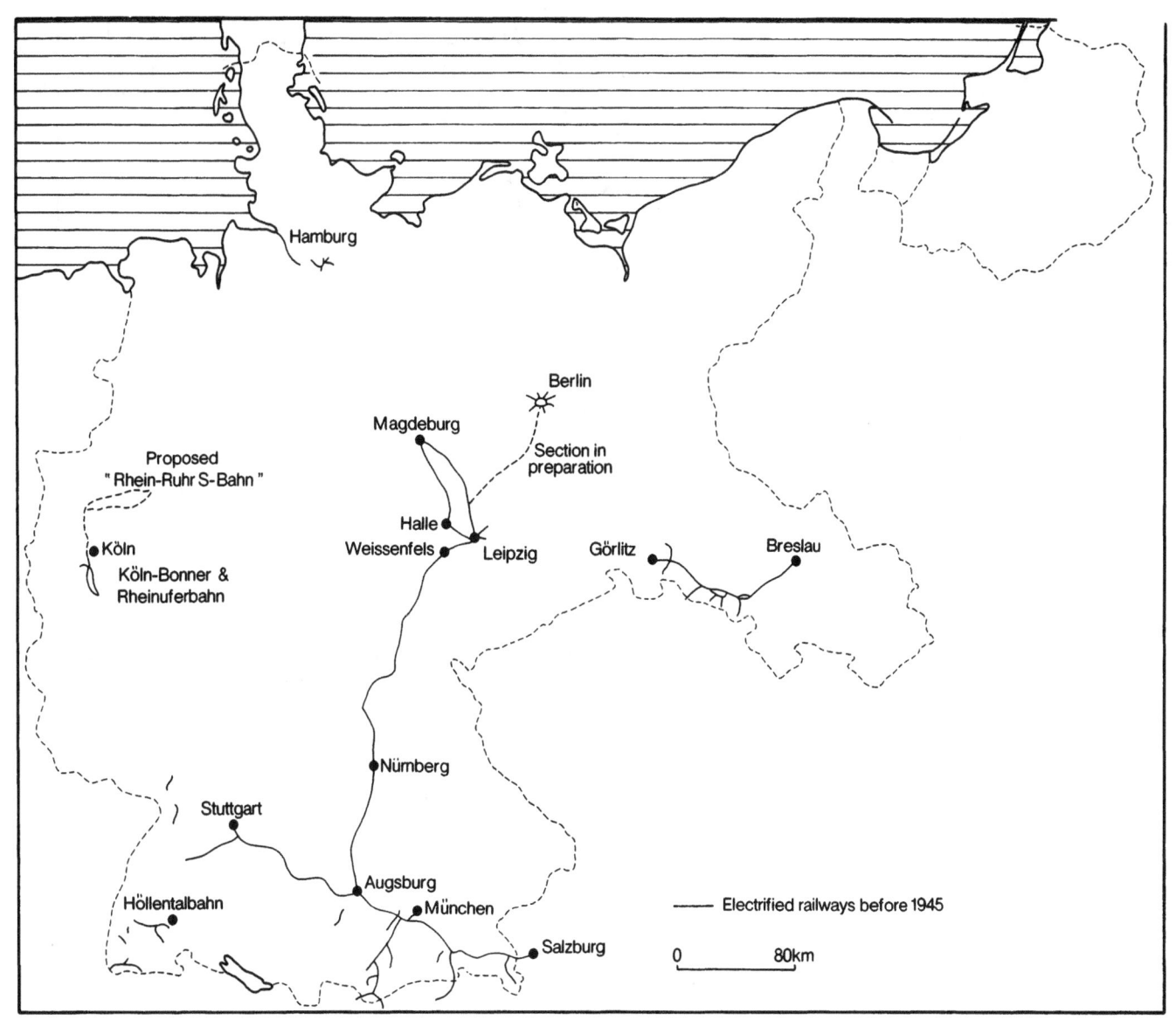

13 Prewar electrification. The through electrified route München-Berlin was never completed, though preparations were far advanced by 1945. A short electrified line at Peenemünde on the Baltic is not shown.
Source: various.

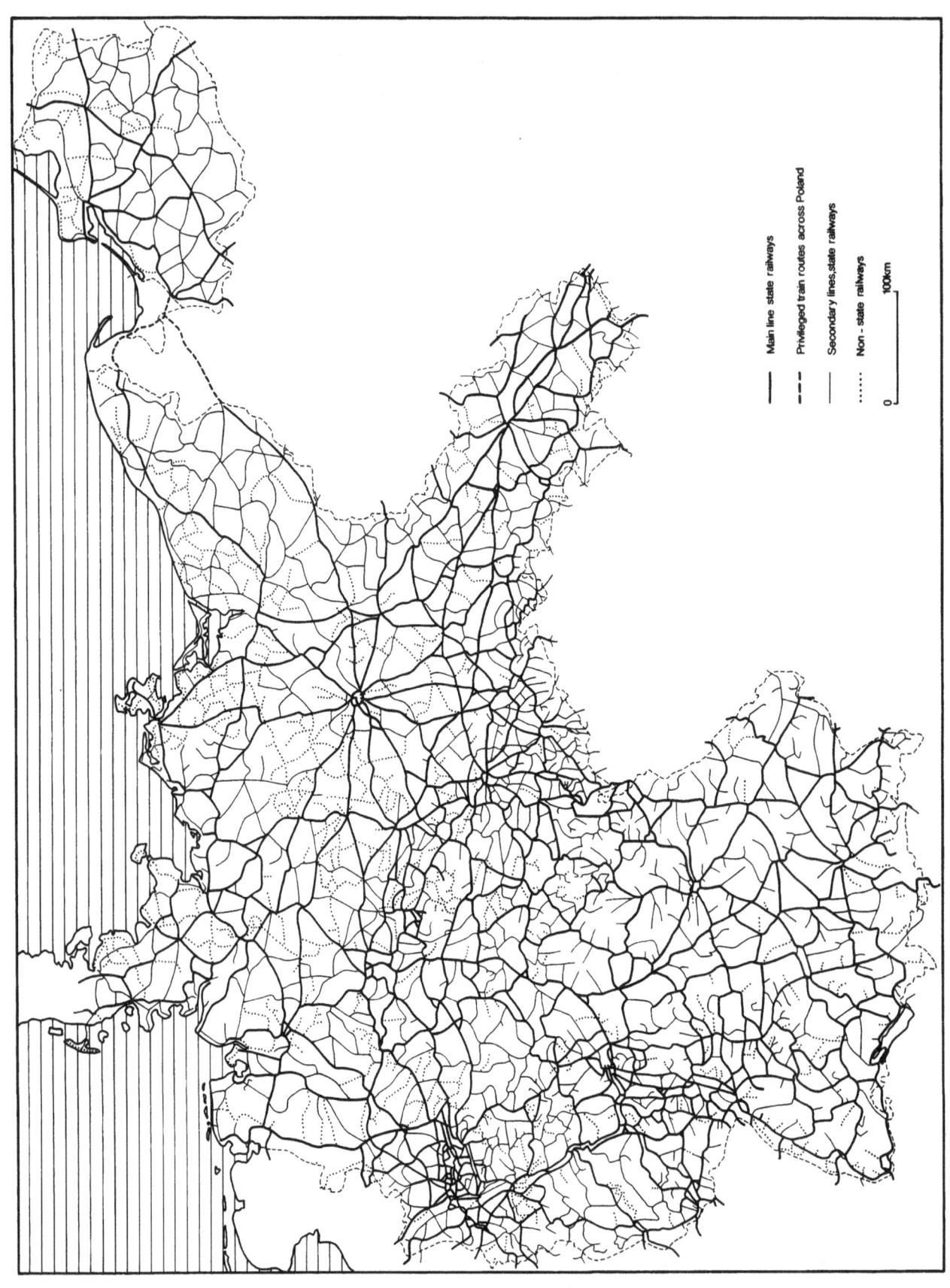

14 <u>Deutsche Reichsbahn</u> network in 1937. The non-state railways shown were mostly lightly loaded rural lines, especially common in the poorer farming districts of Prussia. A few heavily used commuter railways outside state ownership existed, as north of Hamburg and between Köln and Bonn.
<u>Source</u>: <u>Streckenkarte der Deutschen Reichsbahn</u>, <u>Mitropa Kursbuch</u>, Summer 1937, Berlin.

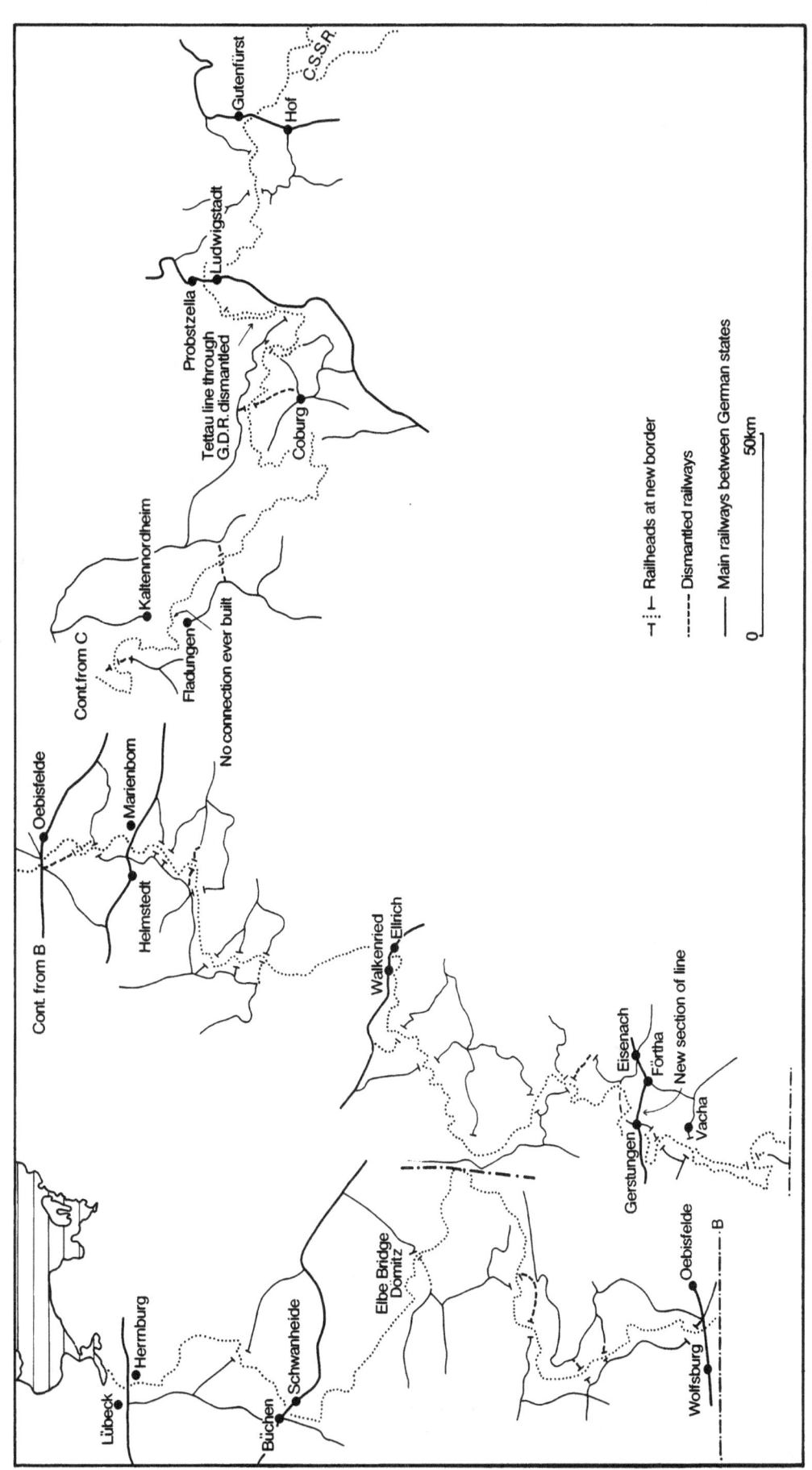

15 Disruption of railways by the post-1945 inter-German frontier. The profound dislocation to east-west routes is apparent from this map. Its full impact can be appreciated by comparison with Fig. 14.

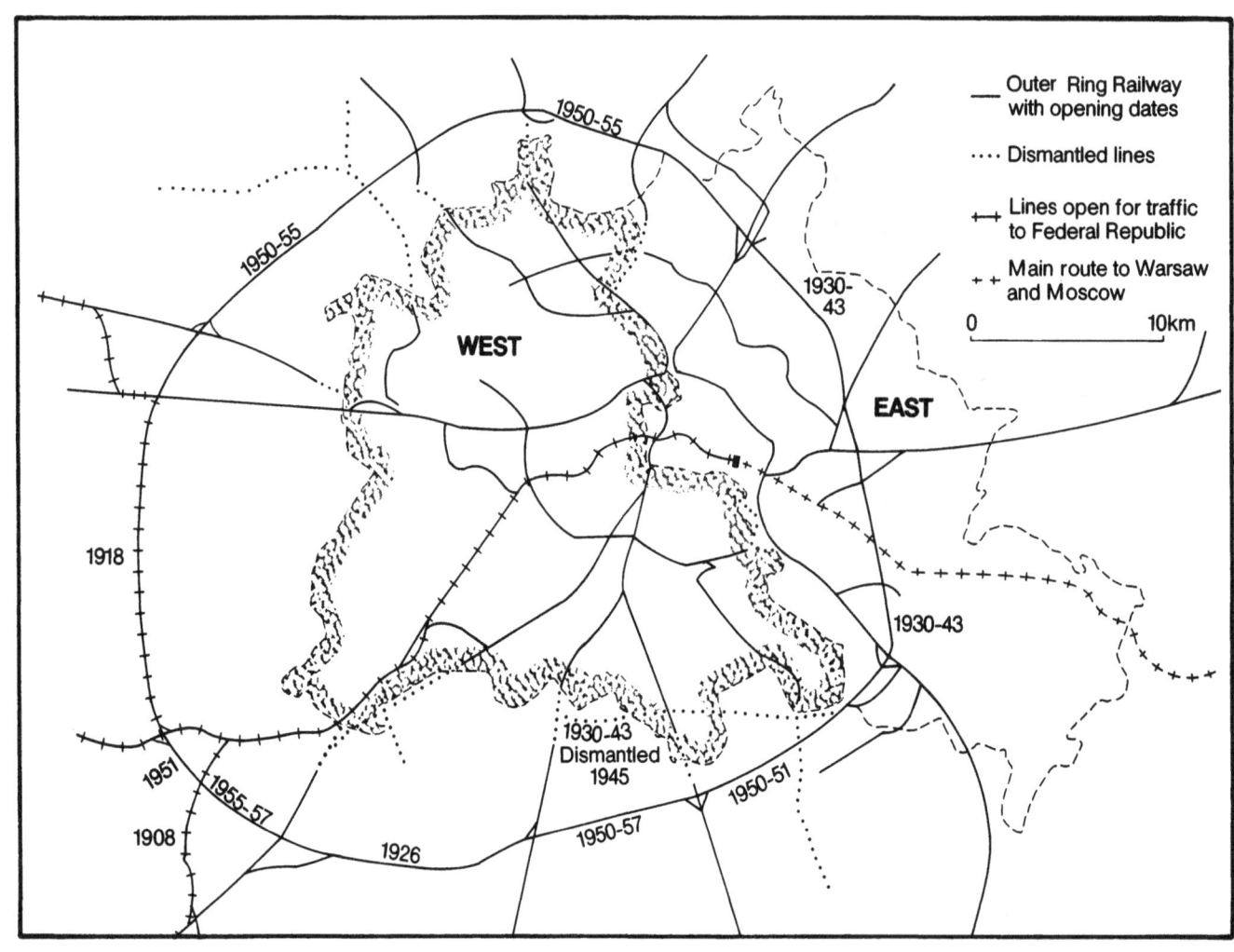

16 Berlin Outer Ring Railway. Completion of this line begun before 1914 has been one of the major construction projects in the German Democratic Republic. It allows all G.D.R. traffic to bypass West Berlin. The map also reflects the disruption in general to railway routes through the partition of Berlin.
Source: various.

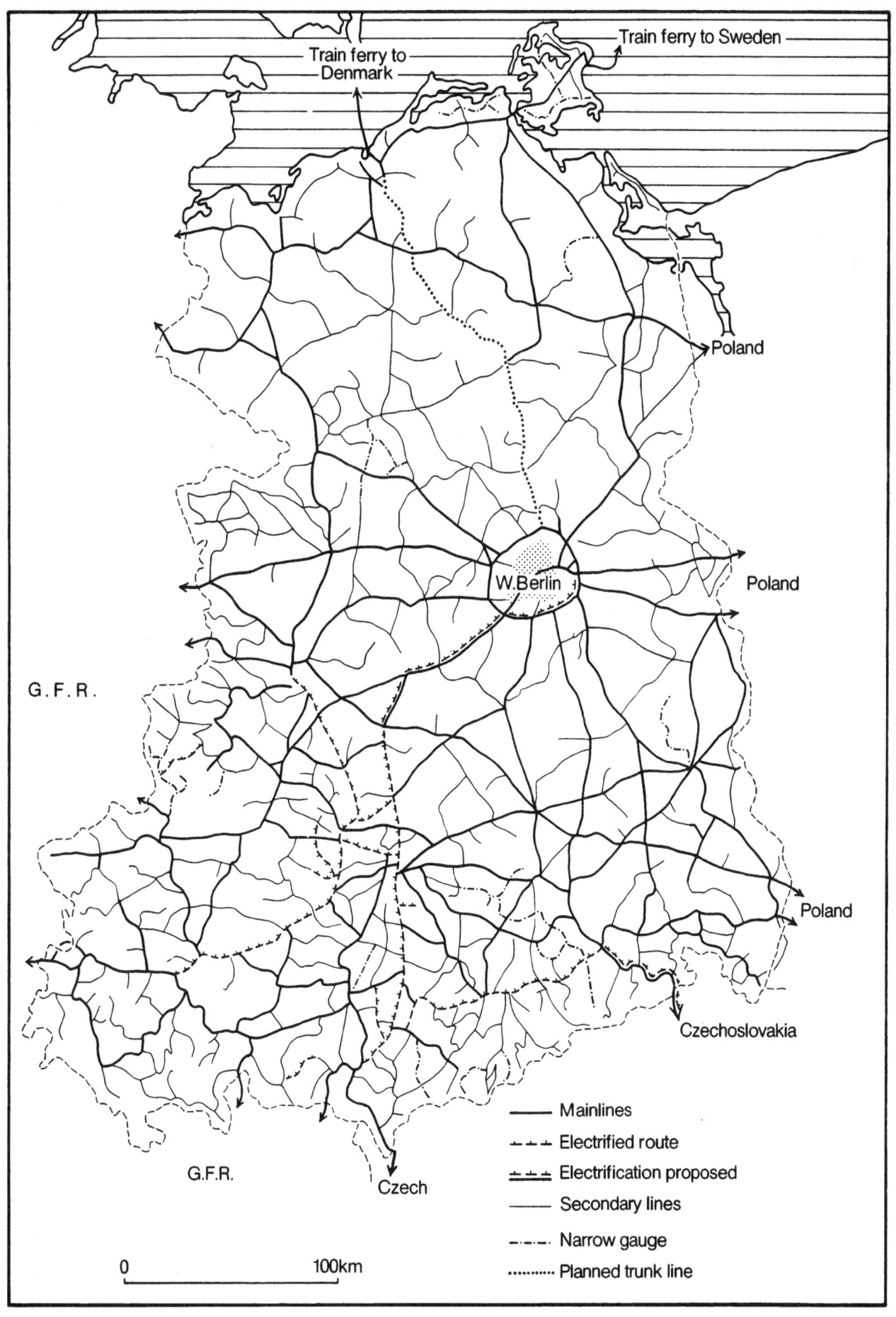

17 Railways in the German Democratic Republic. The planned trunkline Berlin-Rostock is being built largely through improvement of existing local railways. Work has recently begun on the electrification of the main Berlin-Leipzig line.

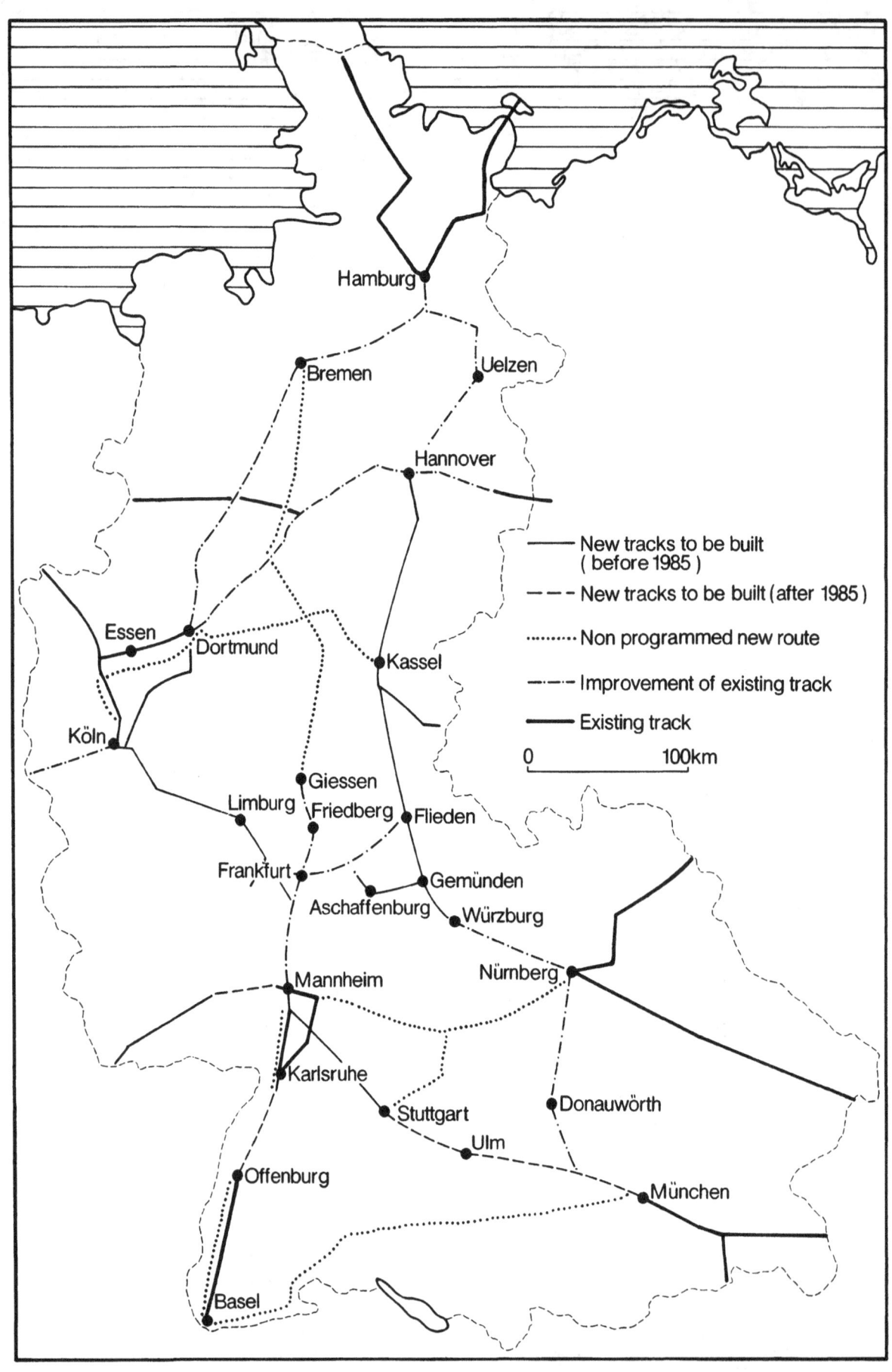

18 Planned new routes in West Germany. Considerable problems of environmental protection have arisen in the design and alignment of these new routes. Priority appears to be on the Karlsruhe-Stuttgart link.
Source: various.

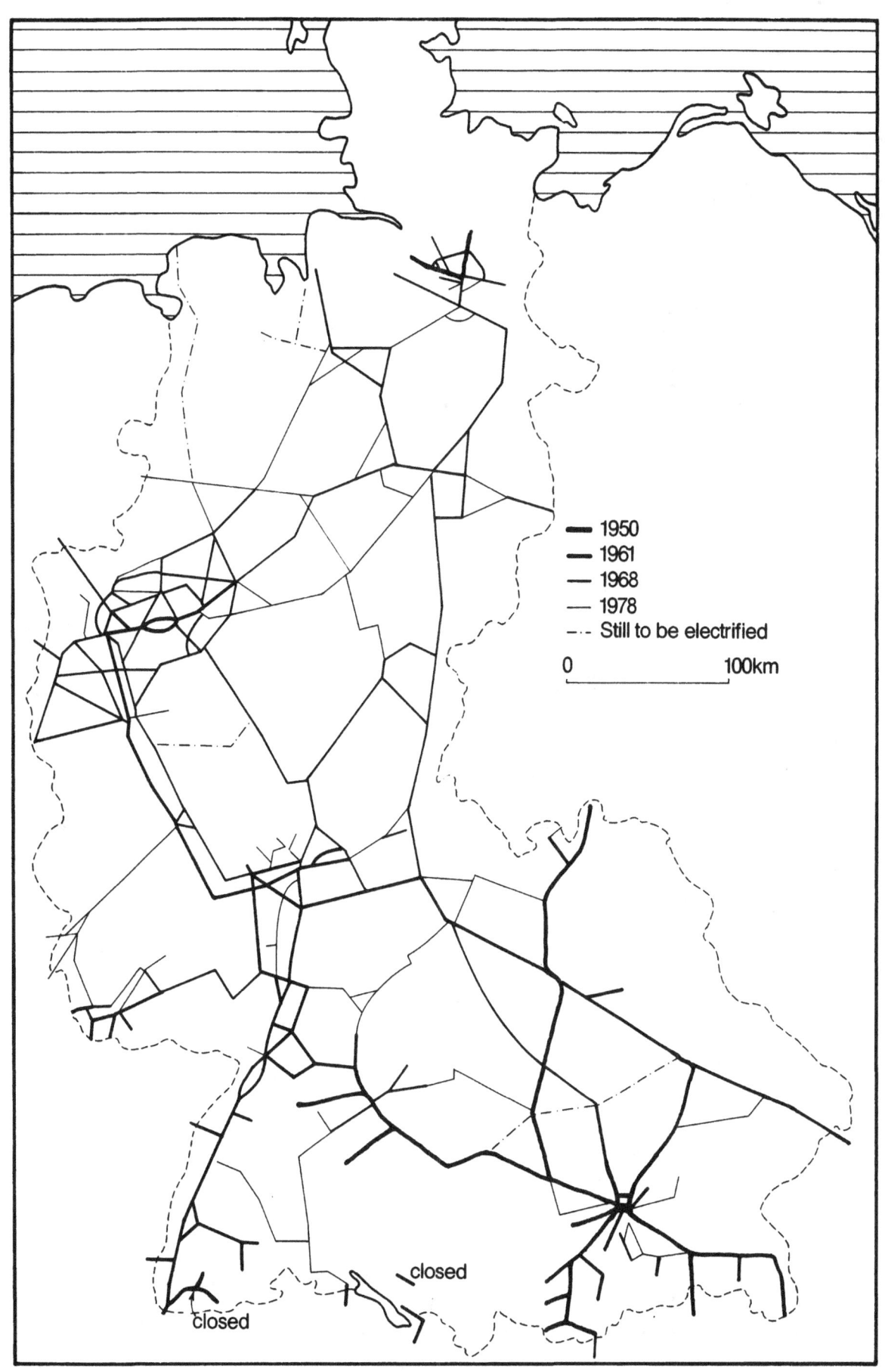

19 Electrification in West Germany. The gradual northwards progression is clear. Note the absence of electrified lines in Schleswig-Holstein. Source: various.

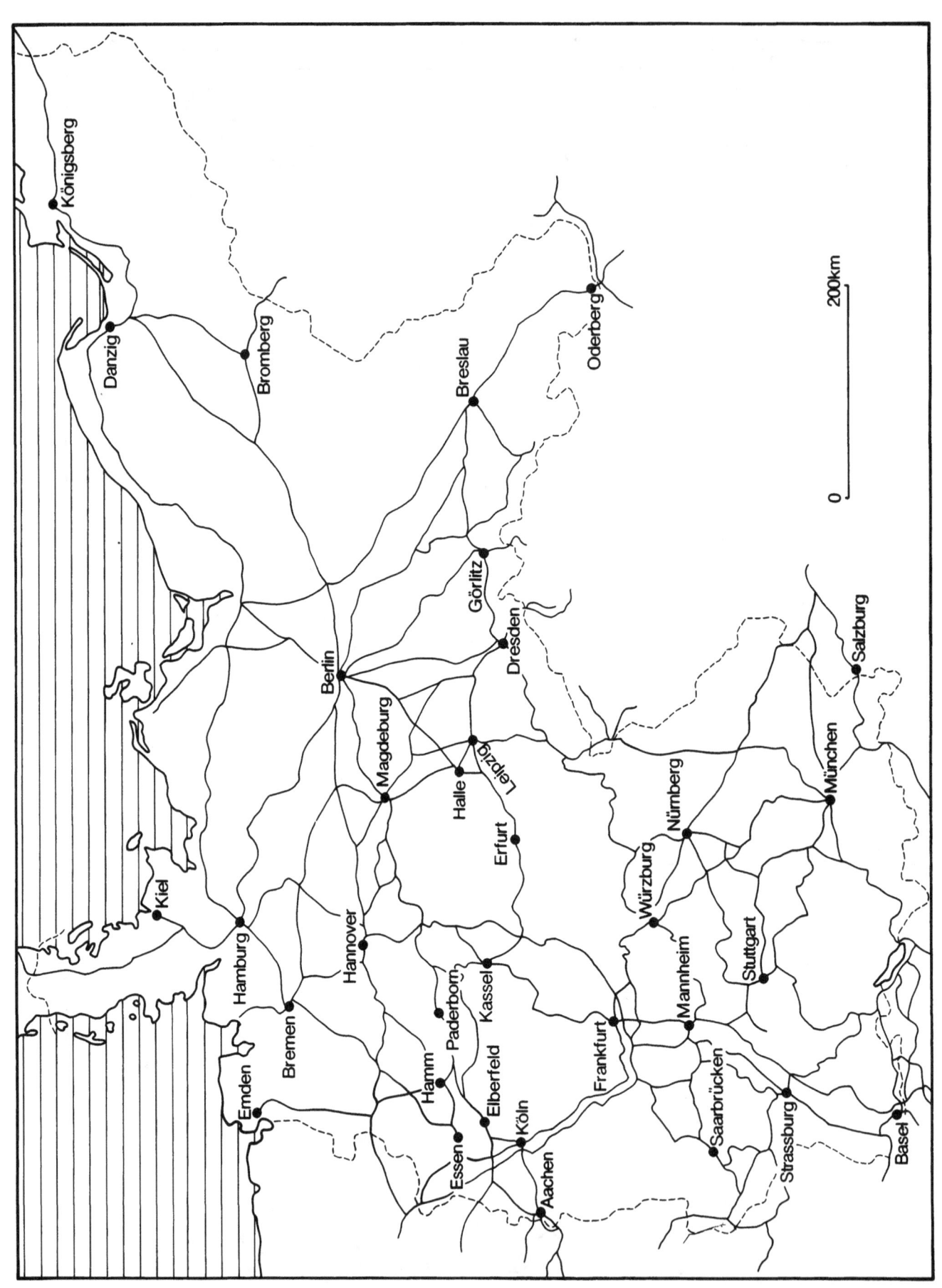

20 Courier trains 1878-1880. An extensive system of such express trains ran over most major routes of the north-south and east-west links, but considerable gaps also remained.
Source: various including Petermanns Geographische Mitteilungen, March 1878, Fig. 10.

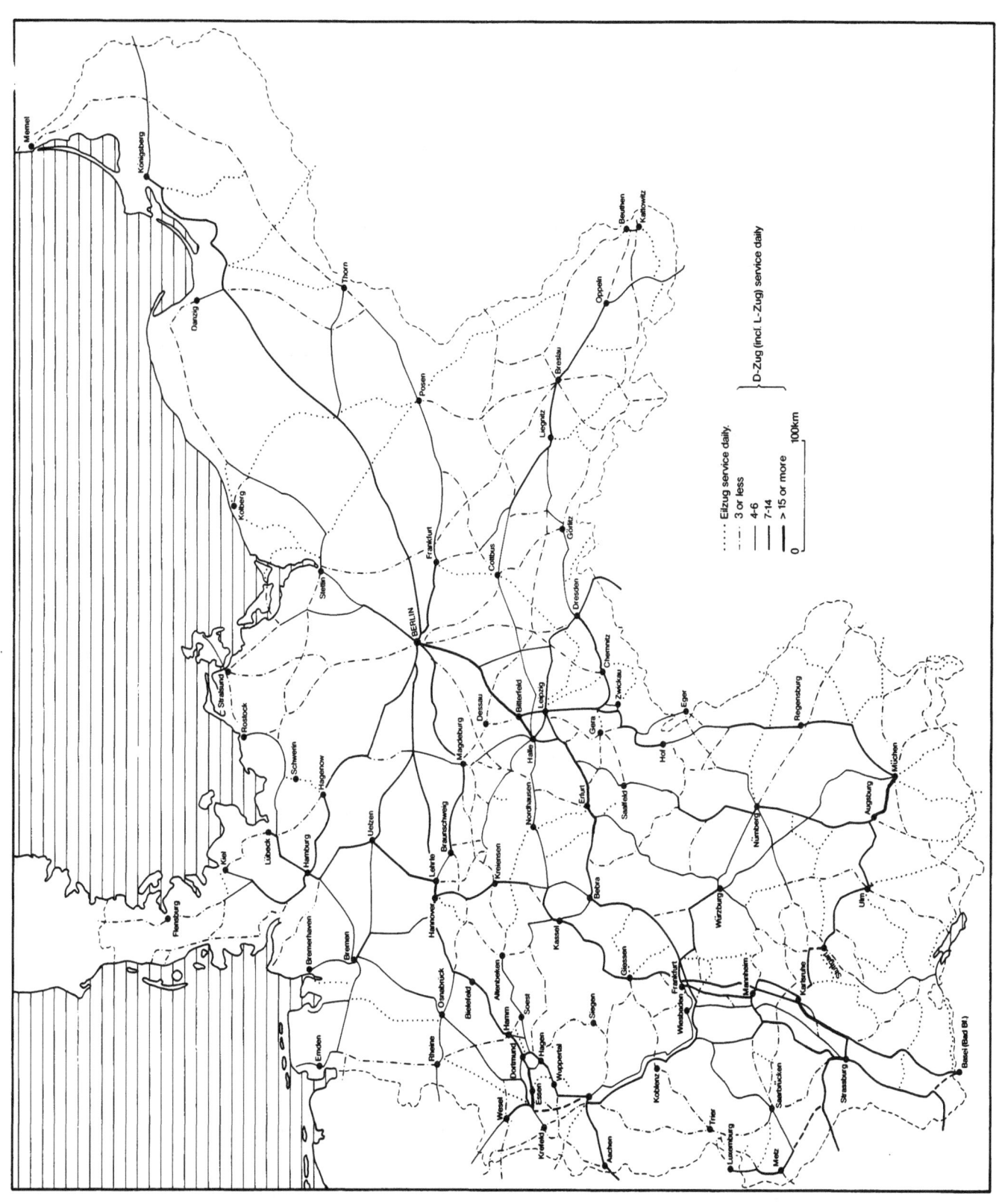

21 Fast passenger train service - Summer 1914. The high level of passenger train services is clear from the variety of routes served and the density of daily connections.
Source: Reichskursbuch, Summer 1914.

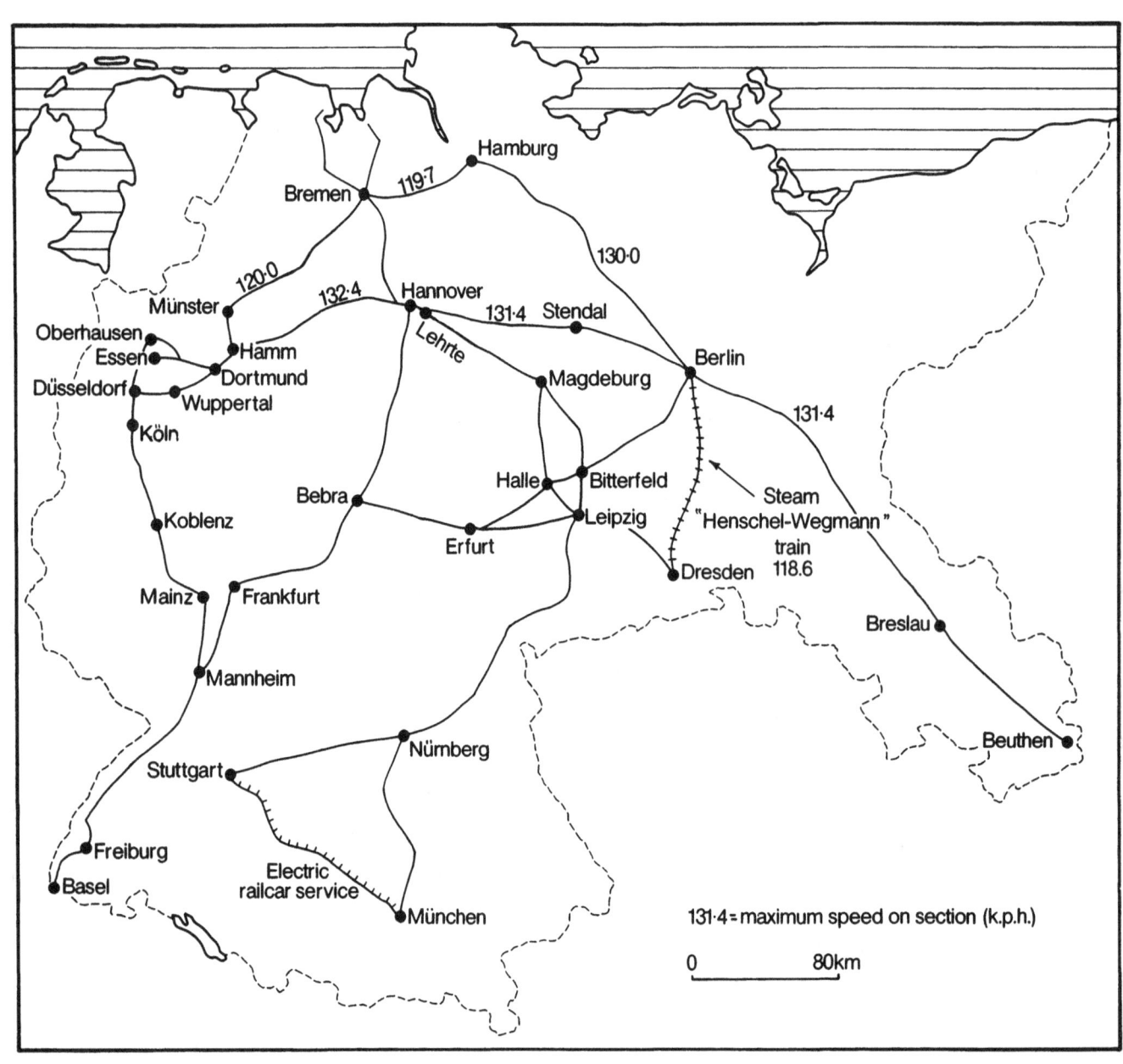

22 Express railcar services – Summer 1939. Some of the highest speeds are indicated. It was possible to make the return journey from Berlin to most provincial centres inside a working day. Where not otherwise shown, services were by diesel railcar sets of the Fliegende Hamburger type.
Source: Reichskursbuch, Summer 1939.

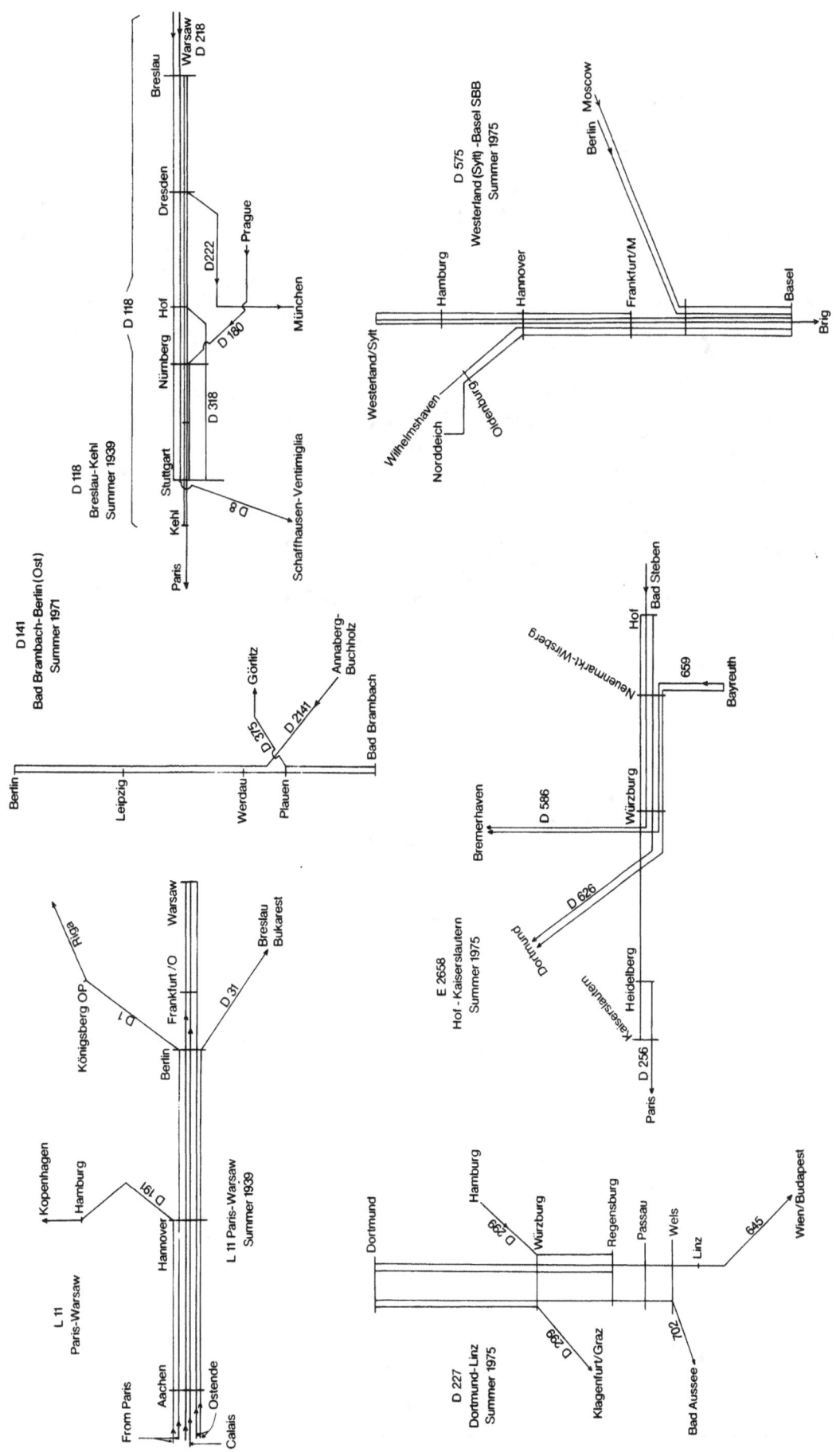

23 Representative through coach services. Through carriages to and from several origins and destinations moving part of their journey in one train have been a common feature of German railway operation. Source: timetables for the respective dates.

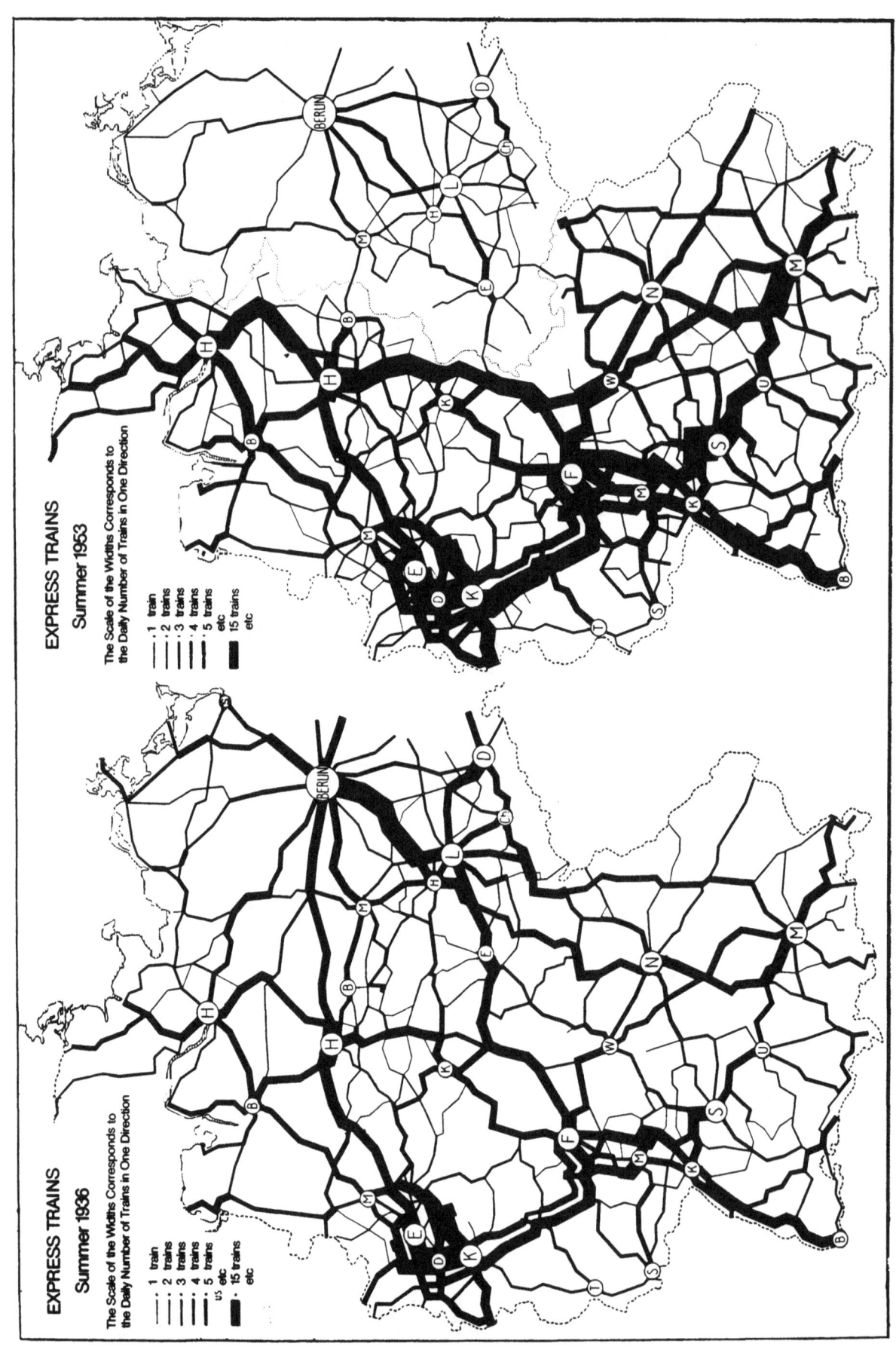

24 Prewar and postwar railway passenger service densities. These two maps highlight vividly the effect of the division of Germany on passenger traffic.
Source: Stöckl, F., *Eisenbahnen in Deutschland*, Heidelberg, 1969.

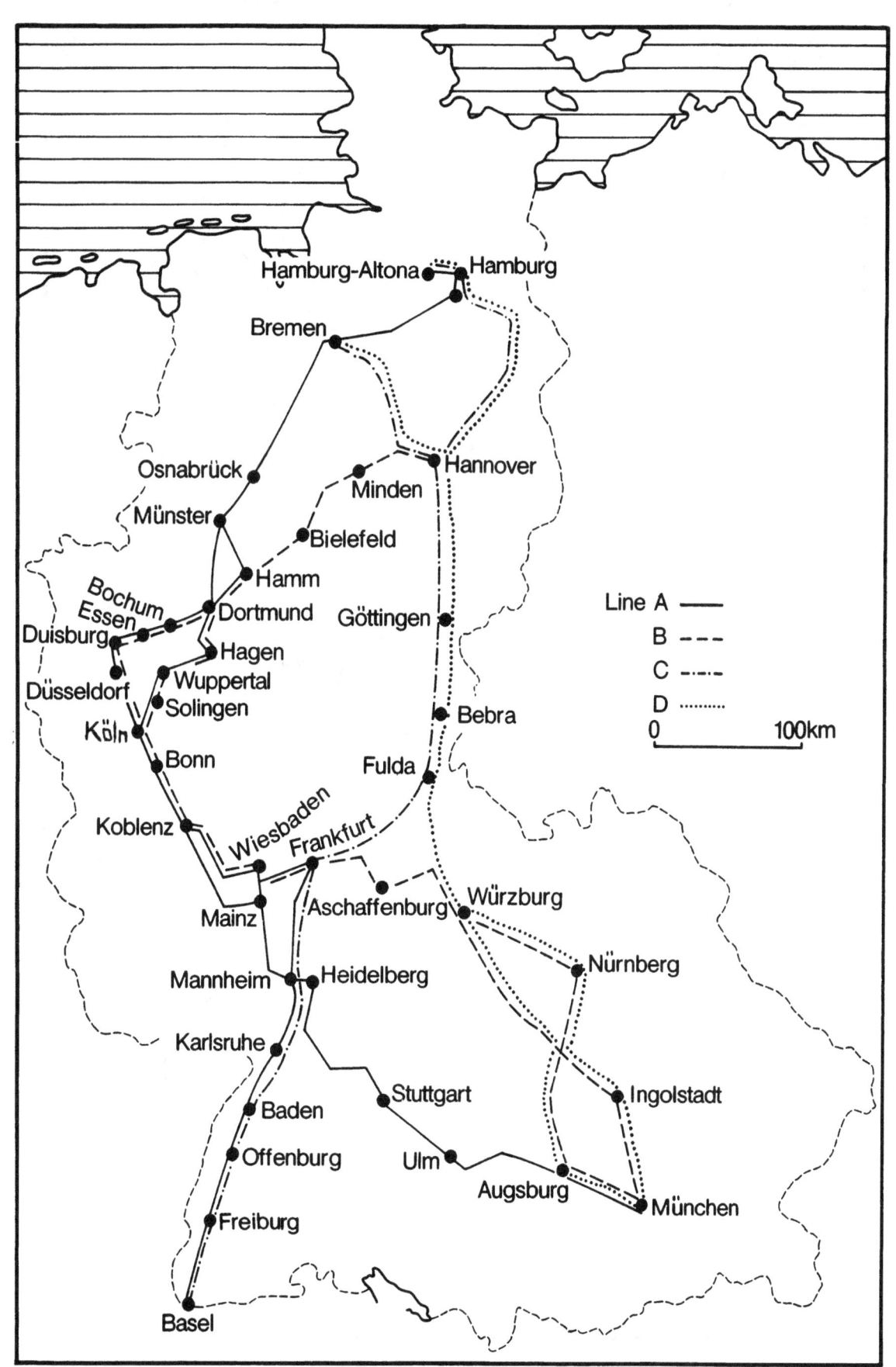

25 <u>Inter-City</u> network in West Germany. These services operate on a near regular interval basis over four main routings (A - D on the figure). Originally first-class only, many of the trains now carry second-class passengers. An elaborate system of feeder trains to these main routings operates.
Source: <u>DB-Kursbuch</u> folder, Summer 1978.

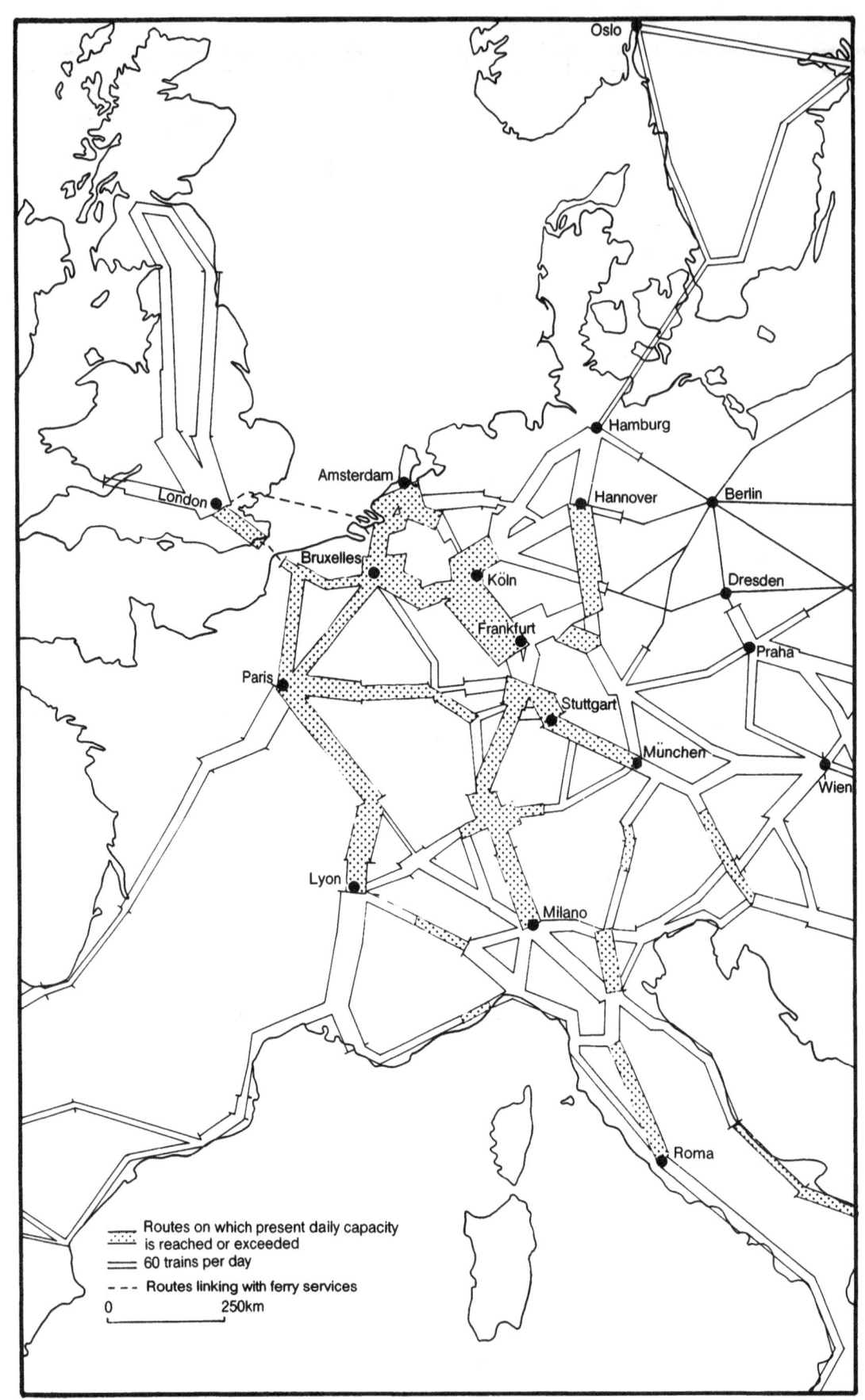

26 Germany in the Union Internationale des Chemins de Fer plan 1973. The U.I.C. seeks to coordinate European railway services overall to compete with air and road competition. The map indicates the routes where expansion of traffic is limited by the capacity of existing facilities.
Source: U.I.C., Plan directeur du chemin de fer européen de l'avenir, Paris 1973.

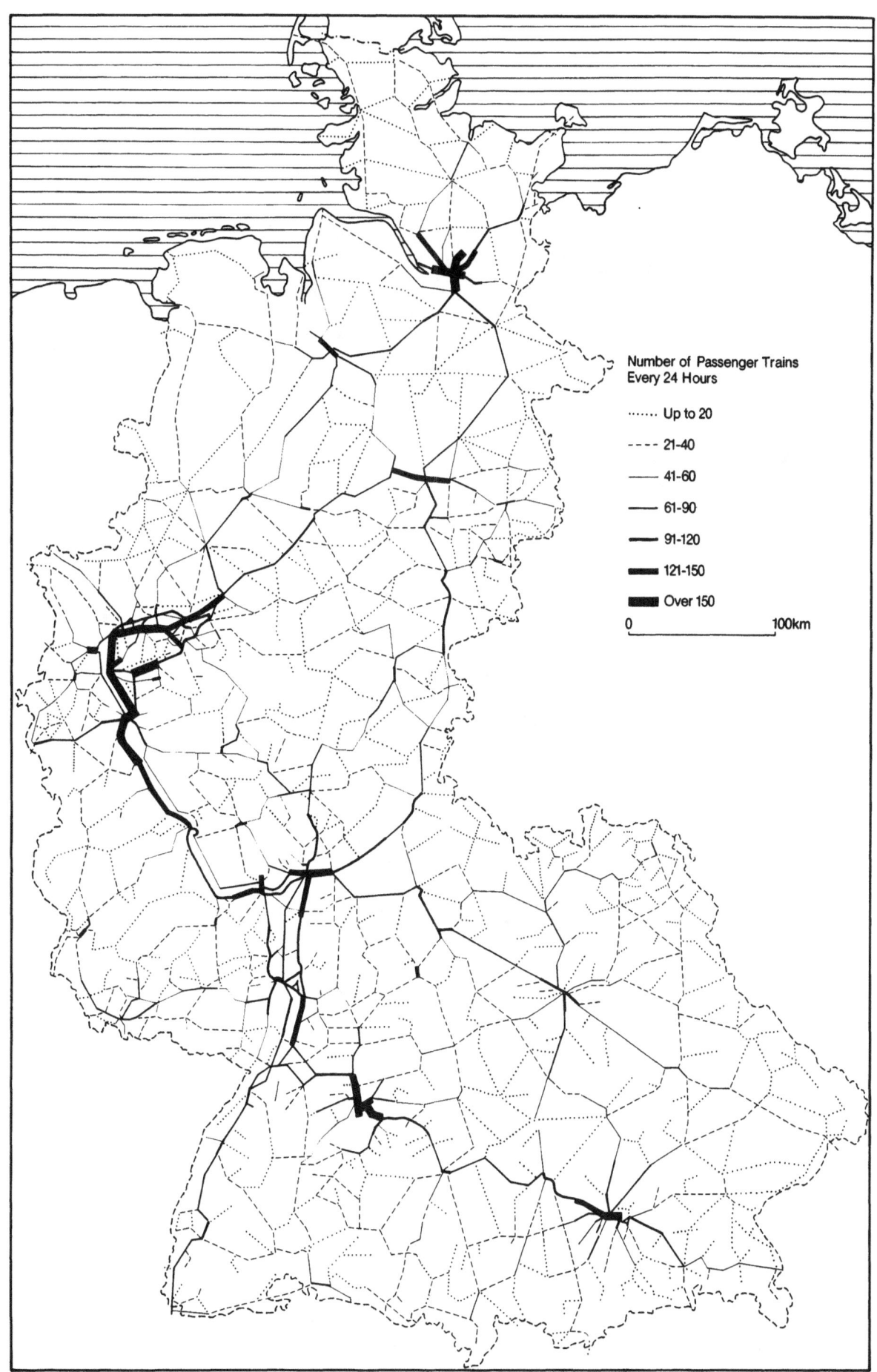

27 Passenger train services in West Germany. All passenger trains are included, so that main commuting areas stand out with high densities.
Source: <u>Atlas der Bundesrepublik Deutschland</u>.

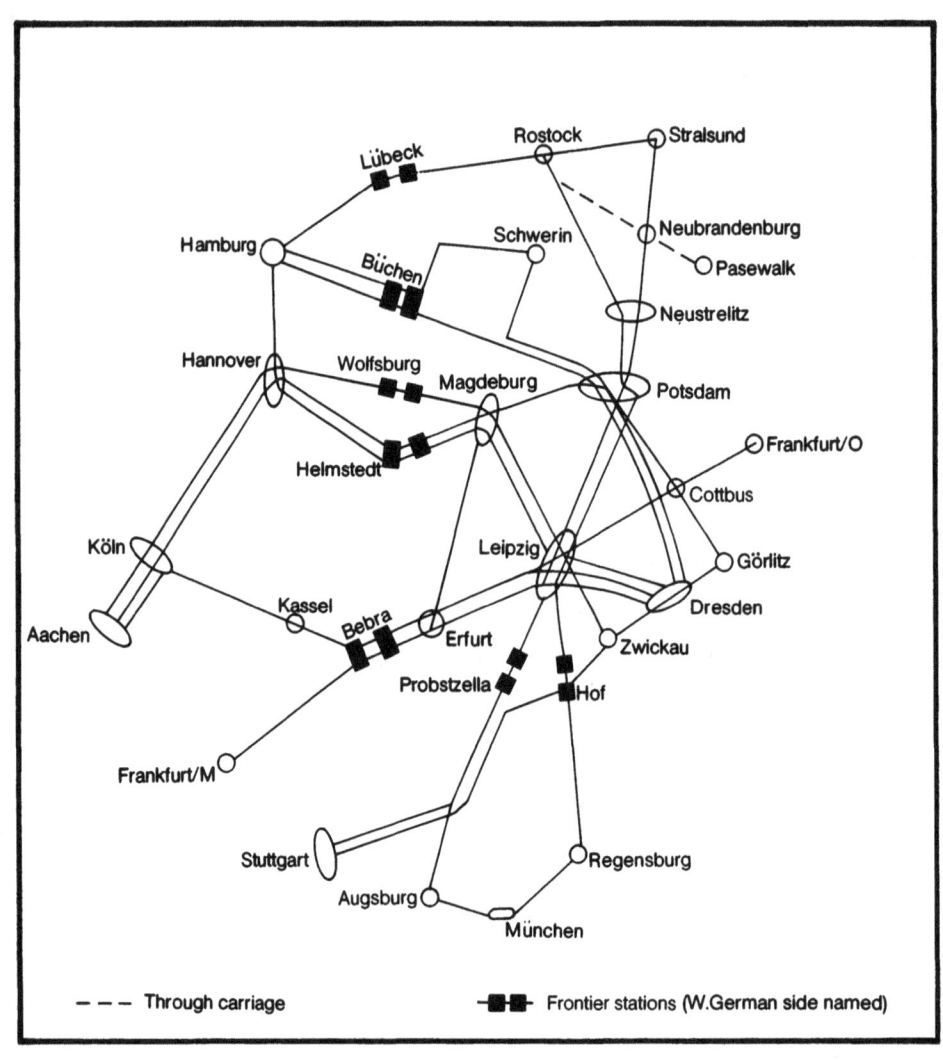

28 Inter-German railway services – Summer 1975. The paucity of services compared to prewar (Fig. 24) is clear. Most routes carry only one or two trains each way daily. West German trains to West Berlin are excluded.
Source: DB-Kursbuch, Summer 1975.

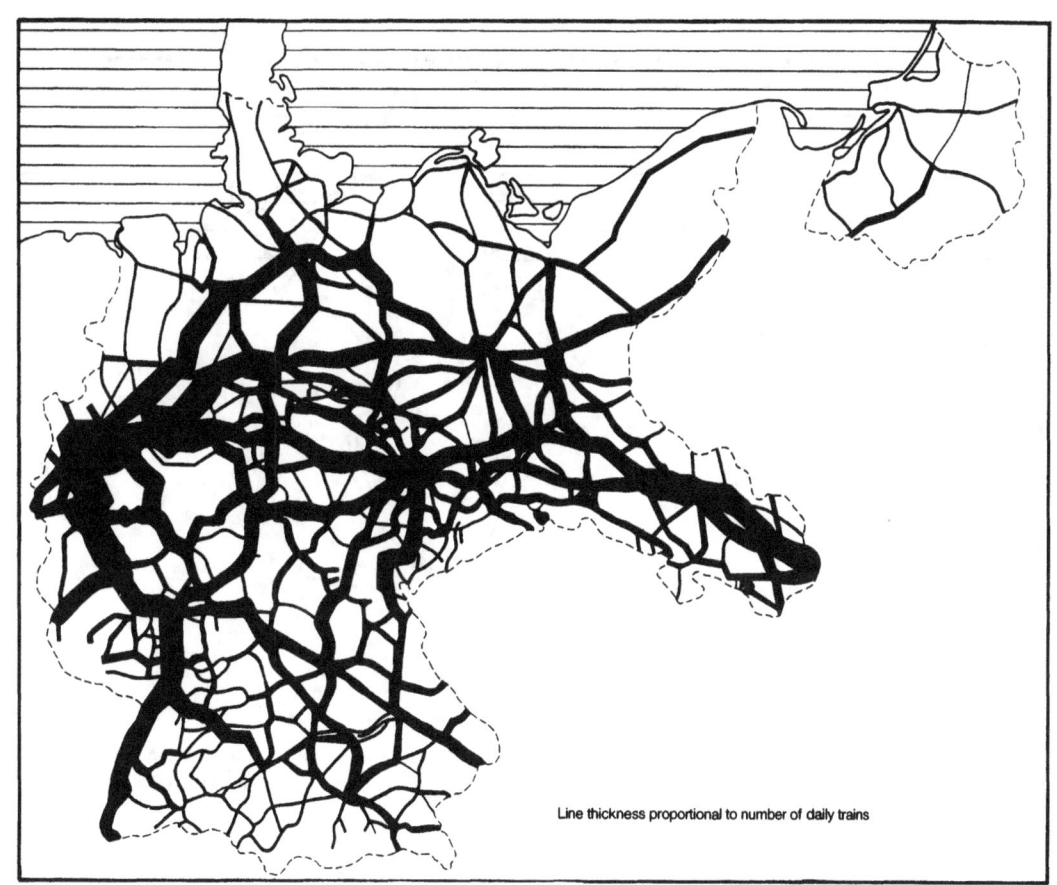

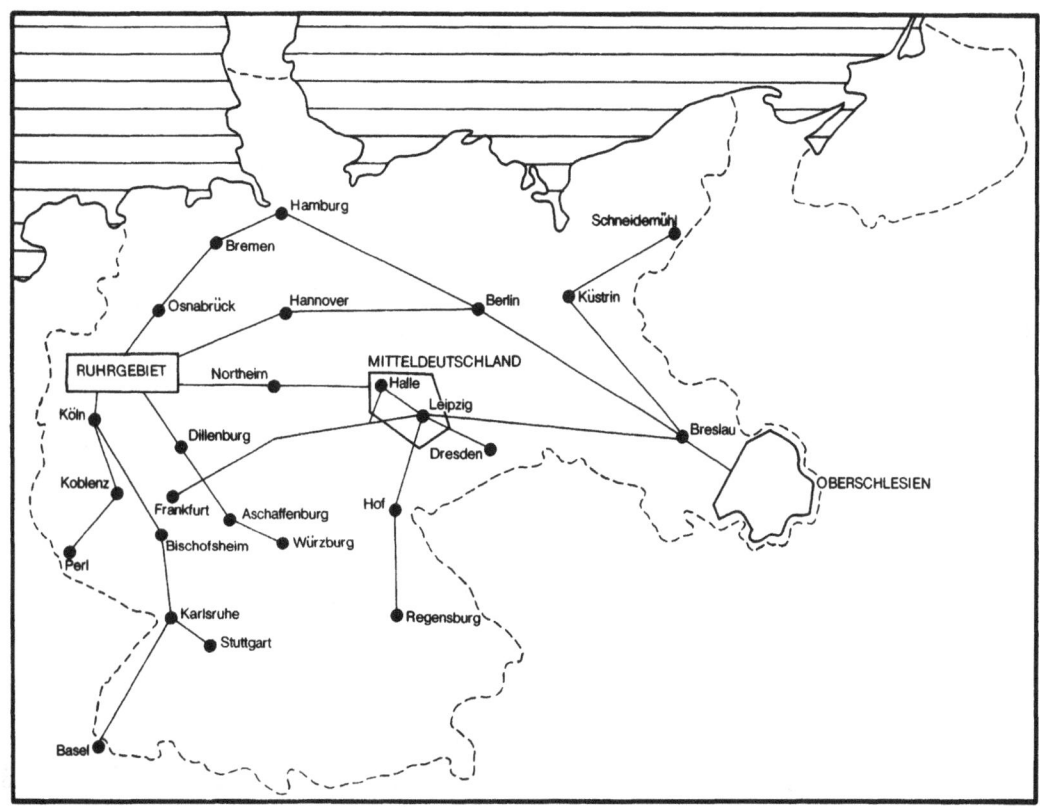

29 Interwar freight movements in Germany. The two maps reflect the importance of the east-west connections between the major heavy industrial districts of the period.
Source: Hundert Jahre Deutsche Eisenbahnen, Berlin 1938, Pfannschmidt, M., Standort, Landesplanung, Baupolitik, Berlin 1932.

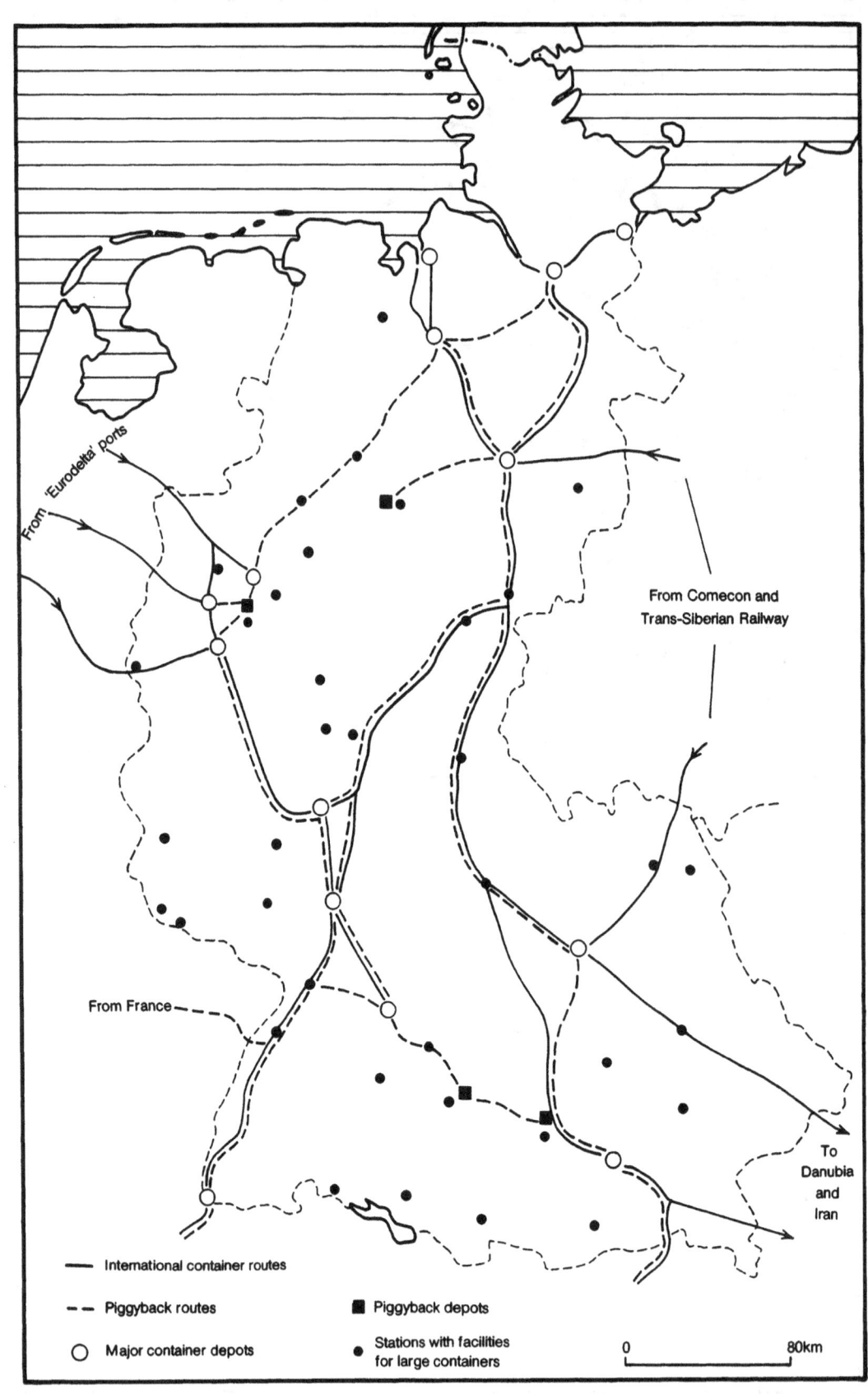

30 Container and piggyback services in West Germany. These new methods of rail freight have quickly established a major network of selected routes.
Source: Marandon, J.C., Der kombinierte Güterverkehr Schiene/Strasse in der Bundesrepublik Deutschland, <u>Materialien zur Raumforschung VI</u>, Bochum 1973.

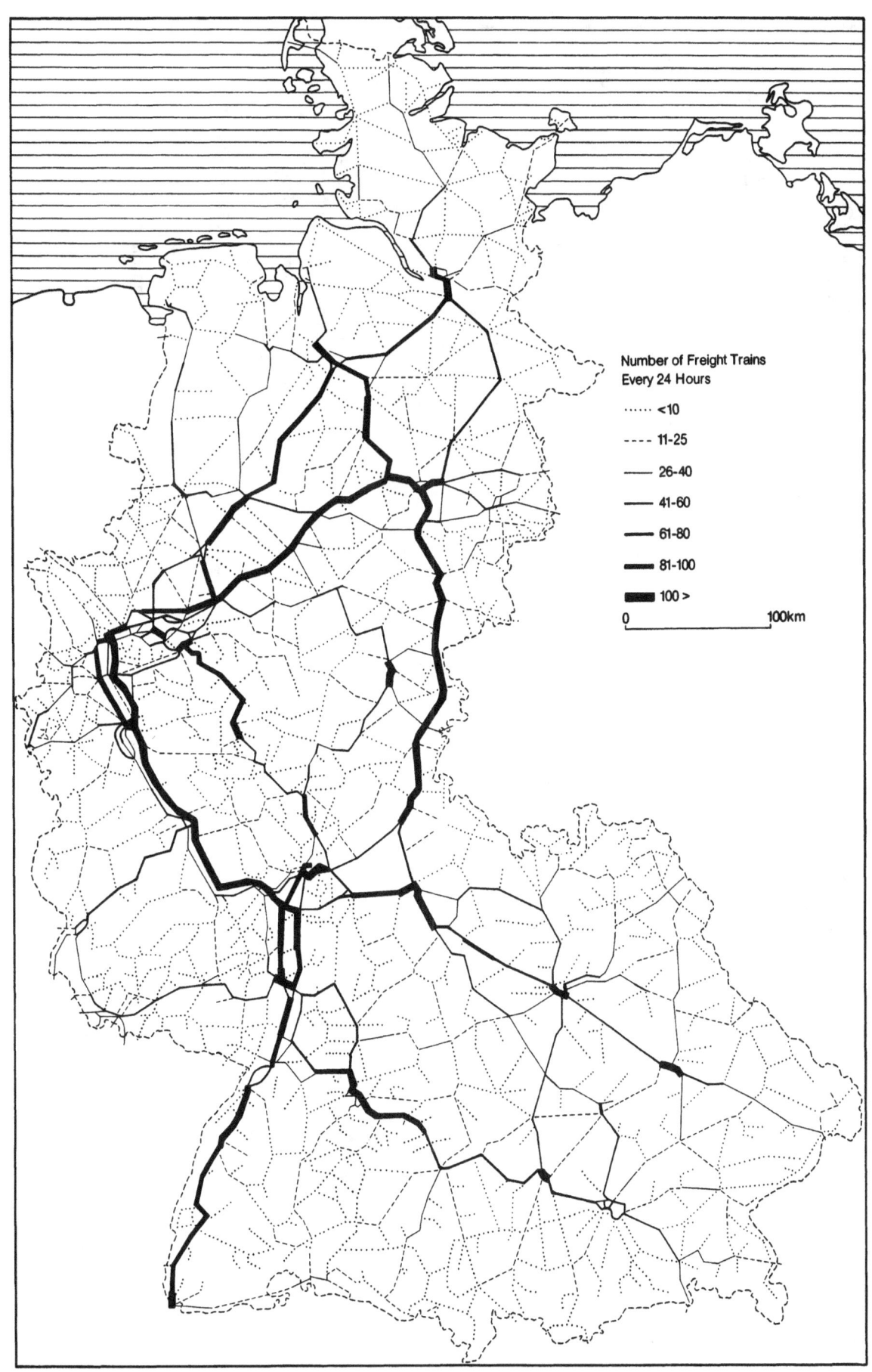

31 Freight services in West Germany. The concentration of freight trains on selected routes is a striking feature.
Source: Atlas der Bundesrepublik Deutschland.

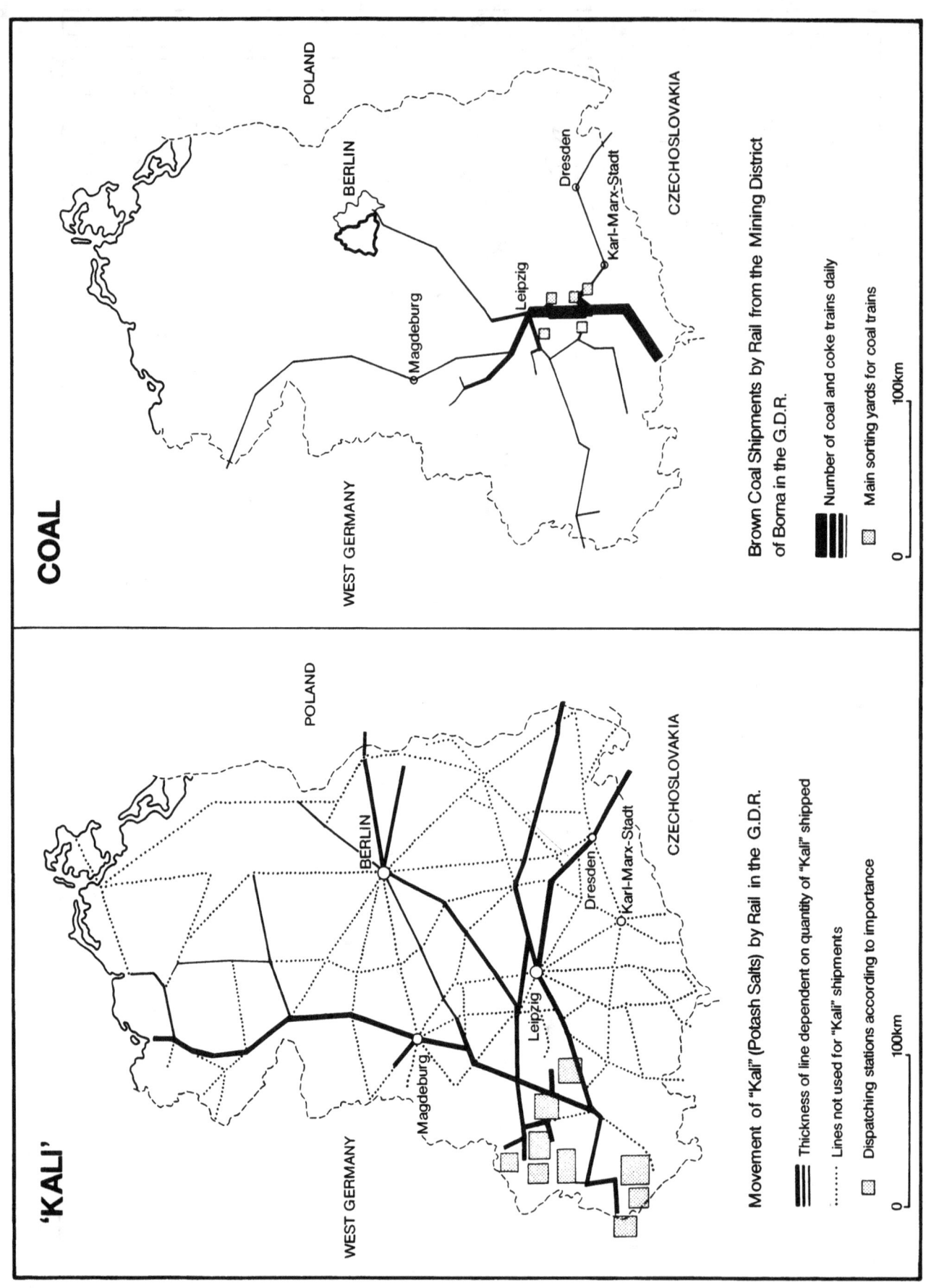

32 Freight services in the German Democratic Republic. Movements of selected commodities reflect the importance of the industrial south and the railways of the so-called Saxon Ring.
Source: Autorenkollektiv, Verkehrsgeographie, Teil 4, Lehrbuch für die......Berufsbildung.....der Deutschen Reichsbahn, Berlin 1968.

For Product Safety Concerns and Information please contact our EU
representative GPSR@taylorandfrancis.com
Taylor & Francis Verlag GmbH, Kaufingerstraße 24, 80331 München, Germany

www.ingramcontent.com/pod-product-compliance
Lightning Source LLC
Chambersburg PA
CBHW082357010526
44113CB00039B/2362